Lecture Notes in Earth Sciences 77

Editors:
S. Bhattacharji, Brooklyn
G. M. Friedman, Brooklyn and Troy
H. J. Neugebauer, Bonn
A. Seilacher, Tuebingen and Yale

W0055349

Springer-Verlag Berlin Heidelberg GmbH

Christian Goltz

Fractal and Chaotic
Properties
of Earthquakes

Springer

Author

Dr. Christian Goltz
Kiel University, Institute of Geophysics
Leibnizstraße 15, D-24118 Kiel, Germany
E-mail: goltz@physik.uni-kiel.de

"For all Lecture Notes in Earth Sciences published till now please see final pages of the book"

Cataloging-in-Publication data applied for

Die Deutsche Bibliothek - CIP-Einheitsaufnahme

Goltz, Christian:
Fractal and chaotic properties of earthquakes / Christian Goltz.

(Lecture notes in earth sciences ; 77)
ISBN 978-3-540-64893-2   ISBN 978-3-540-68459-6 (eBook)
DOI 10.1007/978-3-540-68459-6

ISSN 0930-0317
ISBN 978-3-540-64893-2

Originally published by Springer-Verlag Berlin Heidelberg New York in 1997

Typesetting: Camera ready by author
SPIN: 10677150   32/3142-543210 - Printed on acid-free paper

Table of Contents

List of Figures

List of Tables

List of Tables

1. Introduction

Earthquake prediction research has been carried out for over 100 years now. There has been no success in the sense of a reproducible prediction of the time, location and magnitude of earthquakes (e.g. [Gel97]). The discussion whether earthquakes are predictable at all has gained some impetus during the last few years (e.g. [Tur97, GJKM97, Wys97]). At the same time there has been a great development in the fields of fractal geometry (e.g. [Kor92, HS93]) and nonlinear dynamics (e.g. [ABST93]), especially in numerical analysis. Fractality of real-world earthquake statistics has by now been established beyond doubt (e.g. [Tak90]) and is in agreement with modern models of seismicity (e.g. [Tur97]). While there is no proof of deterministic chaos in real earthquake data, nobody questions nonlinearity of the earthquake process (e.g. [Mei94]) and slider block models have been shown to exhibit chaotic behaviour (e.g. [HT92]).

Have inappropriate (Euclidean, linear) methods of analysis been used so that possibly existing earthquake precursors simply couldn't be detected? Or is it that individual earthquakes are inherently unpredictable due to their chaotic dynamics or high "complexity"? Both issues will be addressed in a theoretical as well as empirical fashion in this book.

Part I discusses the application of fractal concepts to seismicity, while Part II applies ideas from nonlinear analysis. Emphasis is on numerical analysis of real-world data with theoretical background and models introduced where applicable to show the motivation behind the analyses and to aid in the interpretation of the results.

Part I begins with the introduction of fractal fundamentals and discusses the various fractal dimensions including multifractals. Special attention is given to deviations from ideal scaling behaviour and to practical aspects of the numerical determination of fractal properties.

A comprehensive sample application to landslides in Chapter 3 may be read on its own. It deepens the understanding of multifractals and addresses further problems in numerical analysis. Configuration entropy analysis as a promising complementary tool to fractal methods is also applied. Implications of the confirmation of landslides as a multifractal process are discussed.

Chapter 4 summarises the various fractal properties of earthquakes which have been established or rediscovered during the past years.

A fascinating "fractal" property of many natural systems is the Hurst phenomenon which characterises persistence or antipersistence of processes, i.e. their long time memory. This phenomenon is discussed in connection with seismicity in Chapter 5. Hurst analysis is used to describe anisotropy in scaling properties of earthquake fields.

Chapter 6 sheds some light on the relationship between multifractal spectra and phase transitions and what precursory qualities these spectra might have and why.

In Chapter 7 the whole toolbox of fractal analysis is applied to seismicity in a region containing the Kobe earthquake of January 1995. After determination of the overall properties of the earthquake catalogue the temporal variation of fractal properties is obtained to check for fractal precursors. A comparison of overall and aftershock seismicity concludes Part I.

Part II first introduces some principles of nonlinear time series analysis. Analysis of three synthetic time series from different classes of dynamics (quasi-periodic, infinite-dimensional and low-dimensional chaotic) illustrates the methods and shows what may be obtained. Next, nonlinear analysis is applied to radon emission and strain, two prominent earthquake-related real-world time series and to earthquake inter-arrival times directly derived from an earthquake catalogue. Finally, "complexity" of earthquake dynamics is discussed by monitoring the variation of apparent attractor dimension with time.

Part I

Earthquakes and Fractals

2. Fractal Concepts

2.1 Introduction

Mandelbrot [Man77] stated that after his introduction of the concept of fractals, scientists will surely be delighted to be able to describe shapes in a rigorous quantitative fashion they formerly called *grainy, hydralike, in between, pimply, pocky, ramified, seaweedy, strange, tangled, tortuous, wiggly, wispy, wrinkled* and so on.

These adjectives all meant the same thing: The shape was not *Euclidean*—it could not be described by line segments or any elements of standard geometry. The latter is the reason why objects occurring in nature were historically regarded as "imperfect" and mathematical objects with this property (e.g., the Cantor set or the related Devil's Staircase, cf. [Bak86] and section 5.2) were called "monsters" and disregarded as curiosities.

The fundamental mathematical definition of a fractal [Man83] is not useful for practical application (it is, on purpose, not complete either, see for example [Fed88]) as it is based on the *Hausdorff Besicovitch dimension* which is impractical to calculate. Barnsley (1988) describes this in detail (see also section 2.2). In the following, D is used as a generic term for fractal dimensions (of which there are infinitely many), d for the classical (Euclidean) integer dimension (also called linear dimension).

More practical definitions of a fractal are:

- Every set with a noninteger D is a fractal.
  This definition anticipates the definition of the fractal dimension D in section 2.2. It is the most practical in the analysis of experimental data as will be seen.
- Most fractals are invariant under scaling transformations.
  Those invariant under ordinary geometric similarity ("magnification") are called *self-similar*. Another word for self-similarity is *scale invariance*. Scale invariance or self-similarity is the rule, the geometric regularity, behind seemingly unlimited complexity and therefore the only means to characterise a fractal structure. The fractal dimension quantifies the scaling behaviour, i.e. is a quantitative measure of fractal quality. For natural objects, this does not mean that different magnifications of the set can be precisely superimposed but they can be made superposable in a statistical sense.

Figure 2.1 implies the practical consequence of this property: Is the object a mountain or a rock? Only the inclusion of an object with a *characteristic length* (the hammer) enables us to tell. This leads to another way of saying the above: Fractals do not possess a characteristic length: further and further magnified, they never become smooth but stay complex.

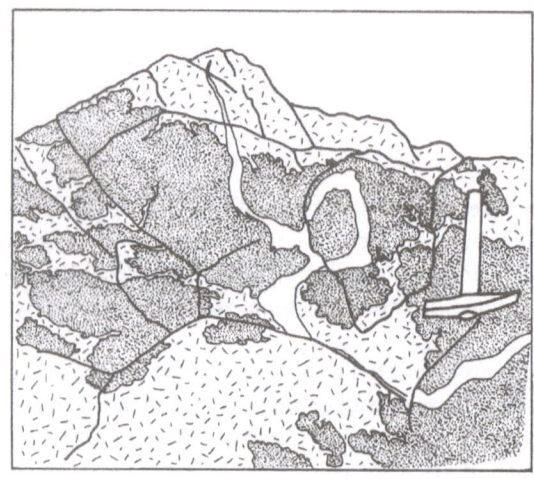

Fig. 2.1. Mountain or rock? An example for scale-invariance (From Bébien *et al.* (1987))

Everyday examples of such objects are smoke, clouds, trees, blood vessels and coastlines. Objective of Part I of this work is to show and describe the fractal structure of earthquakes as for their distribution in space and time and to try and extract possibly useful information for earthquake prediction from their temporal variation. Chapter 4 will summarise some previous findings in this direction.

— Fractal curves are nowhere differentiable.
This is a revolutionary aspect of fractal geometry as it denies the use of ordinary differentiation.
— A power law relation exists between some linear distance *r* and the "mass" (any measure such as seismic energy, size of landslides or number of events) of a fractal.

Several deviations from such ideal fractal properties exist. They will be addressed in section 2.2.7. Leading to the next section: "The only thing which must be studied closely is the fractal dimension"[Tak90].

2.2 Some Important Fractal Dimensions

The definitions of "fractal" and "fractal dimension" go together. Therefore it is not surprising that a definition of *the* fractal dimension does **not** exist.

In fact there is an infinite number of fractal dimensions. Furthermore, even when the type of fractal dimension is given, the method and details of estimation must be mentioned. Otherwise results are not comparable and therefore meaningless.

Often the term "Hausdorff dimension" is used inappropriately for practical estimations of fractal dimensions or specific definitions of *one* fractal definition are called *the* fractal dimension. This is a result of Mandelbrot's "vague" definition of a fractal which was done in order not to exclude future developments and discoveries in the field of fractals.

In the following, only the fractal dimensions which have been applied in this work will be discussed in detail and some related definitions and methods will be mentioned briefly to back up the definitions and methods used later. A similar introduction with less mathematical and numerical detail may be found in [Gol96]. An overview with thorough mathematical background and detailed discussion of the advantages and drawbacks of most fractal dimensions and their methods of determination may be found in [Cut93].

2.2.1 Euclidean and Similarity Dimension

The dimensions of a line, plane and cube are 1, 2 and 3 respectively. In physics, 4-dimensional space-time is common through the addition of a time axis, earthquakes occupy a five-dimensional space: three spatial coordinates, time and size. The integer values for these dimensions coincide with the *degree of freedom*—the number of independent variables (see also Part II). This positive integer dimension is the *Empirical* or *Euclidean Dimension d*.

Dividing a line segment, a square and a cube into similar forms of half the size[1] gives 2^1, 2^2 and 2^3 smaller objects respectively. Taking the exponent to be d, this is in accordance with the Euclidean dimension. Generally, if an object consists of n^d similar shapes of size $1/n$, d is the dimension. When a shape consists of b similar objects of size $1/a$

$$D_S = \frac{\log b}{\log a} \tag{2.1}$$

gives the *similarity dimension*. It can take non-integer values and is therefore the first *fractal* dimension. However, it can only be applied to strictly self-similar shapes and is therefore useless for natural objects. The term $\log b / \log a$ nevertheless serves to understand the following definitions which would otherwise seem arbitrary.

2.2.2 Hausdorff Dimension

The generalisation towards not strictly self-similar shapes is the *Hausdorff dimension* [Hau19] mentioned earlier. It is defined by a method of *covering*:

---

[1] Size means linear dimension here, i.e. (side)length.

Let $D > 0$ and $\epsilon > 0$ be real numbers. Cover a set S by k spheres[2] whose diameters are smaller than ϵ. Denote the radii of the spheres by $r_1, r_2, ..., r_k$. Then the D-dimensional Hausdorff *measure* is defined by:

$$M_D(S) = \lim_{\epsilon \to 0} \inf_{r_k < \epsilon} \sum_1^k r_i^D \qquad (2.2)$$

The Hausdorff Dimension D_H of the set S is the special value of D where the Hausdorff measure varies from infinity to zero.

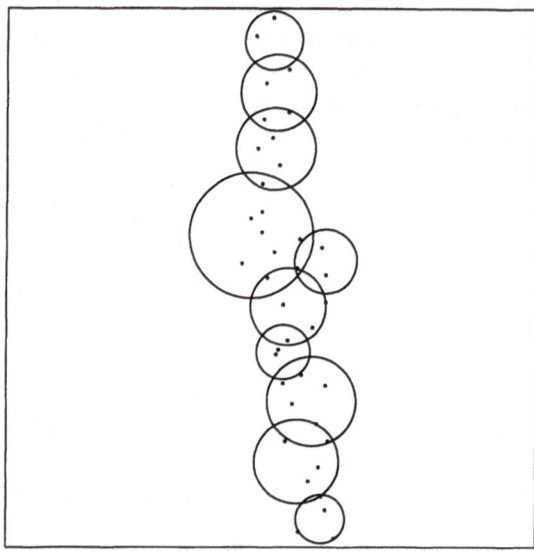

Fig. 2.2. The Hausdorff dimension: the set of points is covered with k overlapping spheres of radii r_i (here $k = 10$) and the Hausdorff measure is calculated by summing the Dth powers of the radii. If the points were samples of a line, the one dimensional measure $(\sum_1^k r_i)$ would give an estimate of the length of the curve. Minimal overlapping of the spheres is guaranteed by taking the infimum in (2.2)

D_H is a generalisation of d and D_S and can be applied to any set of points through the method of covering. Considering the 2-dimensional case, the spheres become circles. These circles may overlap and their radii are not constant as can be seen from Fig. 2.2. From this and the actual definition of D_H it can be seen that not only the mathematical determination but also the estimation by computer is very difficult. Therefore more practically useful definitions of D are needed.

2.2.3 Capacity Dimension D_0

One of them is the *Capacity dimension* D_0. It's practical variant is also called the *box counting dimension* (see below). Like D_H, it is based on the covering of a set by spheres. Kolmogorov [Kol59] introduced the capacity dimension (therefore sometimes also called *Kolmogorov dimension*) to be:

---

[2] A sphere is the surface of a d-dimensional body.

$$D_0 = \lim_{\epsilon \to 0} \frac{\log N(S, \epsilon)}{\log 1/\epsilon} \tag{2.3}$$

where $N(S, \epsilon)$ denotes the smallest number of spheres of size $1/\epsilon$ needed to cover the set S (*minimal covering*).

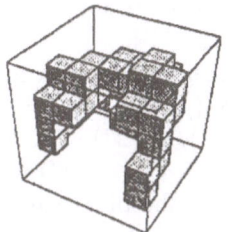 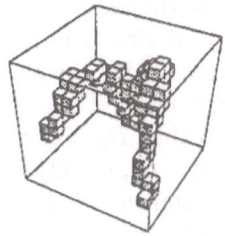 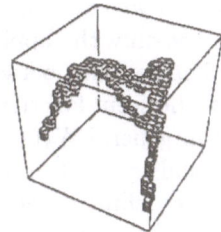

Fig. 2.3. The box-counting dimension: The set of points in three dimensions is covered with spheres ("boxes") of successively smaller size and the number of spheres which contain at least one point is counted at each resolution (after Abarbanel *et al.* (1993))

Definition 2.3 is equivalent to

$$N(S, \epsilon) \propto (1/\epsilon)^{D_0(S)}, \tag{2.4}$$

i.e. a *power law relation* exists between the number of spheres and their size. This power law relation occurs in all (following) definitions of fractal dimensions. In fact this is necessarily so, as there is no other scale-invariant function than a power function (a function $f(x)$ is scale-invariant if $f(x) \propto f(\lambda x)$ for all λ).

D_0 may be seen as a special case of D_H where the sizes ϵ of the spheres are constant. The idea is illustrated in Fig. 2.3. It is

$$D_0 \leq D_H \leq d \tag{2.5}$$

In a sense, D_H describes the size of fractals which possess the same D_0 ([Bar88]). Again, as for all definitions of fractal dimensions, D_0 coincides with all other dimensions in the case of Euclidean shapes (e.g. equality in relation (2.5) is given).

D_0 is a *geometric* measure, it does not account for the frequency of points in the covering spheres, i.e. a possibly non-uniform distribution which might be caused by clustering. Thus an important aspect of the finer structure of a fractal distribution is lost.

Practical Determination of D_0. Replacing ϵ by a discrete variable and using the Euclidean metric (see 2.11), it follows the Box-Counting-Theorem, i.e. the definition of the box counting dimension ([Bar88]):

$$D_0' = \lim_{n \to \infty} \frac{\log N_n(S)}{\log(2^n)} \tag{2.6}$$

for the 2-dimensional case. Then $N_n(S)$ denotes the number of closed adjacent squares of sidelength $1/2^n$ needed to cover the set S in the plane (or: $N_n(S)$ is the number of closed, adjacent squares of sidelength $1/2^n$ which contain at least one point of the set S). This definition is the basis for every computer program which estimates D_0. In the following, D_0 will be used instead of D_0' because only the applied point of view is of interest here.

The equality between d and D_0 for the case of a Euclidean set can now easily be seen: For the line segment (/) it is $N_1(/) = 2$, $N_2(/) = 4$, $N_3(/) = 8$, and in general $N_n(/) = 2^n$, so that $D_0 = \lim_{n \to \infty} \log(2^n)/\log(2^n) = 1$. For the solid square ($\square$) it is $N_n(\square) = 4^n$, i.e. $D_0 = \lim_{n \to \infty} \log(4^n)/\log(2^n) = 2$.

The definition of the box counting dimension leads to the practical *box counting method* (see also [Gol90]):

From the condition $n \to \infty$ in 2.6 it may already be seen that D_0 can only be *estimated* by computer (another, even more critical, reason is that the infinite set S can only be sampled discretely for digital analysis). However, given the finite representation of S, it is covered by successively finer meshes according to 2.6 and the number of non-empty boxes is counted for each resolution.

Then $\log N_n(S)$ is plotted against $\log(2^n)$ and a line is fitted through these points. The slope of this line approximates D_0. Usually a least squares fit, i.e. a maximum likelihood approach, is used to carry out this linear regression. Obviously, the standard deviation of the error, which describes the quality of the fit, can be used as a first simple measure for the reliability of the estimate. As will be seen later, however, the standard deviation is not sufficient to estimate the error in D (not to mention the coefficient of correlation; cf. [GMM98]). Also other methods, such as taking for example half the difference of the maximum and minimum piecewise slope over a fixed range in the $\log - \log$ plots (cf. [SR95]) are problematic; in general, no method one could automatically rely on exists. Goltz (1996) has used non-linear optimisation for the automatic simultaneous determination of scaling region and exponent but found that manual control is still required sometimes. Appendix A outlines a modified algorithm which is very fast and thus preferable for large data sets in low embedding dimensions.

2.2.4 Information Dimension D_1

Extending the definition of D_0 leads to the *information dimension D_1* which is especially applicable to stochastic, i.e. uncorrelated, distributions of points. Before giving it's definition, some background from information theory is useful (cf. also [Gol96]): The amount of information (or surprise) associated with the occurrence of an event E with probability $P(E)$ is measured by

$$I(E) = -\log P(E) = \log 1/P(E). \tag{2.7}$$

This becomes clear when noting that if $P(E) = 1$, i.e. when the event is certain, $I(E) = 0$, i.e. there is no surprise at all. On the contrary, if $P(E) = 0$, $I(E)$ as the limit of $P(E) \to 0$ becomes infinite—the surprise about the impossible happening is unlimited. Furthermore this measure has the property $I(E) = I(F) + I(G)$ when F and G are two independent events and the occurrence of E is dependant on the simultaneous occurrence of F and G. It means that the surprise about two independent events happening simultaneously is equal to the sum of the information conveyed by each event individually.

Again covering the set S with a minimal number $N(S, \epsilon)$ of spheres of size ϵ, the probability $P_i(S, \epsilon)$ for finding a randomly chosen point in the ith sphere can be calculated. Now the information conveyed by each cell (the surprise of finding a point in sphere i) is $I_i(S, \epsilon) = -\log P_i(S, \epsilon)$. The average $(\sum_i P_i(S, \epsilon) = 1)$

$$I(S, \epsilon) = \sum_{i=1}^{N(S,\epsilon)} -P_i(S, \epsilon) \log P_i(S, \epsilon) \qquad (2.8)$$

measures the average information conveyed by finding which sphere a point is in[3]. This quantity is also called the information entropy (entropy in the sense of information theory, see for example [Sha81]).

Together with the shown properties of (2.7), it becomes clear that $I(S, \epsilon)$ is also a measure for unpredictability: If the points are uniformly distributed in S, $I(S, \epsilon)$ has it's maximum (minimum predictability). In the extreme non-uniform case (all points are clustered in one sphere) it is $P_1 = 1, P_{i \neq 1} = 0$ and therefore $I = 1 \log 1 = 0$ (maximum predictability).

Defining a fractal dimension to be:

$$D_1 = \lim_{\epsilon \to 0} \frac{I(S, \epsilon)}{\log 1/\epsilon} \qquad (2.9)$$

finally gives the information dimension. This definition means that if $I(S, \epsilon) \propto -D_1 \log \epsilon$ with varying ϵ, D_1 is a fractal dimension.

The similarity between (2.3) and (2.9) shows that D_1 is a generalisation of D_0: If all probabilities are equal $(P_i(S, \epsilon) = 1/N(S, \epsilon))$, i.e. the distribution of points is uniform, (2.8) becomes $\log N(S, \epsilon)$ and $D_1 = D_0$. Furthermore, if the points are uniformly distributed in d-dimensional space, D_1 becomes d: In this case $P_i(S, \epsilon)$ simply depends on the sphere size, i.e. $P_i(S, \epsilon) \propto \epsilon^d$. Substituting this into (2.8) the equality follows from (2.9). Thus the definition of D_1 is also in accordance with the intuitive notion of the empirical dimension.

If the data is non-uniform, the information dimension is smaller than the capacity dimension. This leads to

$$D_1 \leq D_0 \qquad (2.10)$$

---

[3] When $\log_2$ is used, $I(S, \epsilon)$ is in bits, which is a practical property for computer application.

Therefore D_1 can be said to quantify the non-uniformity of the point distribution by giving less weight to spheres which contain less points than others. The idea is illustrated in Fig. 2.4.

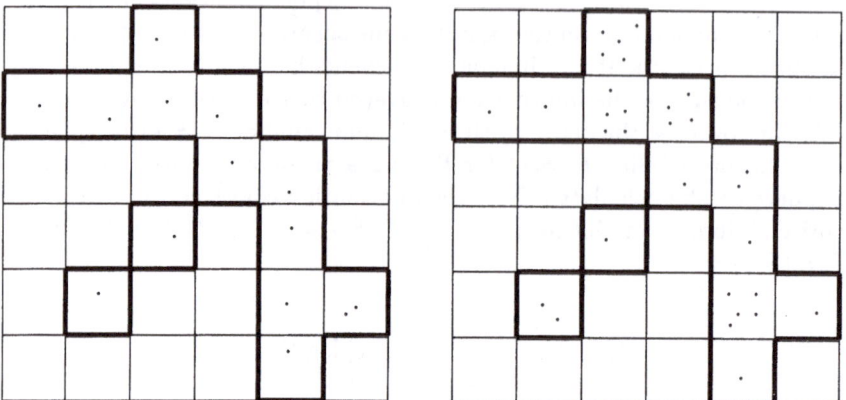

Fig. 2.4. The information dimension: The process of estimating D_1 is shown for one specific stage of the successively finer covering. In the left part the minimal covering with boxes of the minimum possible size is shown: each box contains only one point, the set contains no further information below this resolution. The outline of the fractal (the fractal support), as it is seen by the computer at this resolution, is the same. In the left part, however, the point distribution is uniform, while it is non-uniform in the right part. Therefore D_0 is the same for both situations, while D_1 differs.

Practical Determination of D_1. Analogous to the determination of D_0, the method to estimate D_1 is straightforward: For each minimal covering of the set S (a minimal covering is the one that uses the minimum number of spheres of a given size) the probabilities $P_i(S, \epsilon)$ are calculated to be $N_i(S, \epsilon)/ \parallel S \parallel$ where $\parallel S \parallel$ is the total number of points in S. The limit calculation is replaced by averaging techniques because of the finite sample.

2.2.5 Correlation Dimension D_2

A further generalisation leads to the *correlation dimension* D_2 which is not based on a covering of the regarded set, but on the distances (spatial correlations) between pairs of points of the set (practically, nevertheless, the correlations are determined by a method of covering again, see below):

First the distances between all N distinct points $(\boldsymbol{x_i}, \boldsymbol{x_j}, i \neq j)$ are calculated using for example the Euclidean norm

$$\parallel \boldsymbol{x_i} - \boldsymbol{x_j} \parallel = \sqrt{\sum_{k=1}^{d} \left(x_{i,k} - x_{j,k} \right)^2} \qquad (2.11)$$

or the maximum norm

$$\| \boldsymbol{x_i} - \boldsymbol{x_j} \| = \max_{k=1}^{d}(x_{i,k} - x_{j,k})$$

where d is the dimensionality of the embedding space.

Then the correlation function is defined to be

$$C(r) = \lim_{N \to \infty} \frac{1}{N^2} \sum_{i,j=1, i \neq j}^{N} \Theta(r - \| \boldsymbol{x_i} - \boldsymbol{x_j} \|) \qquad (2.12)$$

Θ is the Heavyside function ($\Theta = 0$ if the argument is less than zero, otherwise $\Theta = 1$). For a fractal set, the correlation function has a power law dependence on r:

$$\lim_{r \to 0} C(r) \propto r^{D_2} \qquad (2.13)$$

Thus

$$D_2 = \lim_{r \to 0} \frac{\log C(r)}{\log r} \qquad (2.14)$$

$C(r)$ accounts for the probability of finding two points in the same sphere ([GP84]) or, in other words, measures the number of points $\boldsymbol{x_j}$ that are correlated with each other in a sphere of radius r around the reference points $\boldsymbol{x_j}$ ([Kru91]).

It can be shown that $D_2 \leq D_1$ and therefore, keeping (2.5) and (2.10) in mind:

$$D_2 \leq D_1 \leq D_0 \leq d \qquad (2.15)$$

As should be clear from the remarks made when discussing D_1, D_2 even more emphasises densely populated spheres as opposed to sparse spheres. The latter generally leads to better statistics, making the correlation dimension the most encountered measure for fractal scaling in the literature when analysing real world data sets.

Practical Determination of D_2. Grassberger and Procaccia (1993) () have shown that $C(r)$ may be more effectively calculated by constructing spheres around fixed points $\boldsymbol{x_i}$ and counting the number of points $\boldsymbol{x_j}$ in these spheres as is shown in Fig. 2.5. This method is called sphere counting and it is realized by the Grassberger-Procaccia-algorithm. Additionally, $C(r)$ is usually only determined for a number of reference points $N_{ref} < N$ to save computation time.

(2.12) thus becomes

$$C(r) = \lim_{r \to 0} \frac{1}{N_{ref}} \frac{1}{N} \sum_{i=1}^{N_{ref}} \sum_{j=1}^{N} \Theta(r - \| \boldsymbol{x_i} - \boldsymbol{x_j} \|). \qquad (2.16)$$

To prevent confusion between the usage of correlation function (2.12) and correlation sum (also called correlation integral), the definition of the correlation dimension is given once more in (2.17) using the correlation sum:

$$D_2 = \lim_{N \to \infty} \lim_{r \to 0} \frac{\log C(r)}{\log r} \qquad (2.17)$$

where

$$C(r) = \frac{1}{N_{ref}} \frac{1}{N} \sum_{i=1}^{N_{ref}} \sum_{j=1}^{N} \Theta(r - \| \, x_i - x_j \, \|). \qquad (2.18)$$

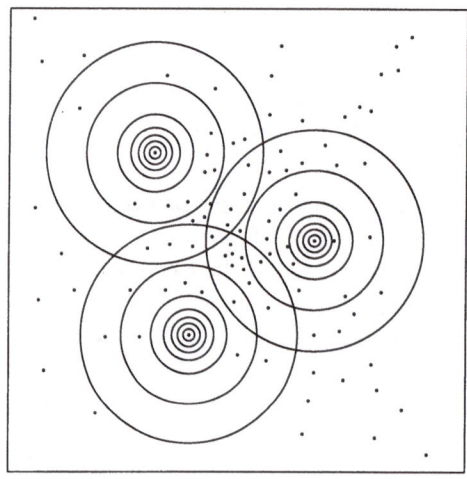

Fig. 2.5. The correlation dimension: The number of points within the varying radii of spheres around fixed reference points is counted. Here seven successive radii for three reference points are shown

As has been mentioned briefly above, the correlation dimension is more accurate for small data sets because it weighs heavier those regions of the embedding space that contain data [GP83, GP84]. Greenside *et al.* [GWSP82] have shown that it is very difficult to obtain a reliable estimation of $D_0 > \approx$ 2: The very large number of data points $N = M^{D_0}$ with $10 \le M \le 42$ has been estimated to be necessary in various cases ([Smi88]). Therefore generally D_2 should be preferred to D_0 if the expected fractal dimension is high ([Smi88, AAD+86, HE89]). A recent result by Hong and Hong (the former from the Seismological Bureau of Chengdu) (1994) is that

$$N_{min} \overset{!}{>} \sqrt{2} \left(\sqrt{27.5} \right)^{D_2},$$

i.e. to be able to obtain a reliable estimate for D_2 in the order of 3, about 204 data points are required at least (when using the sphere counting algorithm). The latter result is much lower than previous ones and makes the determination of D_2 seem feasible for quite small data sets. The correlation dimension belongs to the class of pointwise dimensions in that it is sensitive to the local behaviour of S in the vicinity of a specific point. An estimate for the overall fractal dimension is obtained by averaging. The popular sandbox method ([Vic92]) also belongs to this category. All pointwise estimators are prone to boundary effects because if reference points are chosen too close to the edges of S, the sphere contents will be underestimated.

2.2.6 Generalised Dimensions and Multifractals

Although each of the successive definitions of D_0,D_1 and D_2 is a generalisation of the previous one, it might already be apparent that a single dimension can not entirely characterise a non-uniform (inhomogeneous) fractal distribution. If the fractal distribution possesses different degrees of clustering in different vicinities, but all groups of clusters of equal "density" (more precisely: of equal local fractal dimension) again form fractals, the distribution is called a multifractal. In that sense, a multifractal is an intertwined set of fractals.

A detailed introduction to multifractals from the geoscientific view is given in Goltz (1996) and not to be repeated here. In the context of the previous sections, however, it is interesting to see how all of the above fractal dimensions can be derived from one single formula ([HP83, GP83, Moo92]): Let

$$I(q,\epsilon) = \frac{1}{1-q} \log \sum_{i=1}^{N(\epsilon)} P(i,\epsilon)^q \tag{2.19}$$

where $P(i,\epsilon)^q$ is the qth power of the probability that points of S lie in the ith cell of a minimal covering of S with spheres of size ϵ and $N(\epsilon)$ is the number of spheres (more generally, P can denote any measure on the fractal support).

Then

$$D_q(q) = \lim_{\epsilon \to 0} \frac{I(q,\epsilon)}{\log \epsilon} \tag{2.20}$$

defines the generalised fractal dimension D_q.

As can be easily verified, for $q \to 0$, i.e. harmonic mean, one obtains D_0, for $q \to 1$ we get D_1 and for $q \to 2$, i.e. arithmetic mean, D_2 results.

For a heterogeneous fractal, D_∞ is the lower limit of fractal dimensions, i.e. the fractal dimension of the most intensive clustering, while $D_{-\infty}$ characterises the least intensive clustering within a point distribution (or the lowest concentration of seismic energy etc.).

Generally, the resulting D_q curve for the multifractal case looks like the one given in the left part of Fig. 2.6 while a monofractal produces a straight horizontal line through D_0. Also shown are the parametric curves $f(q)$ and $\alpha(q)$.

Shown in the right part of Fig. 2.6 is the typical $f(\alpha) - \alpha$ curve and its relation with the D_q curve: Simply speaking, the $f(\alpha) - \alpha$ curve describes the

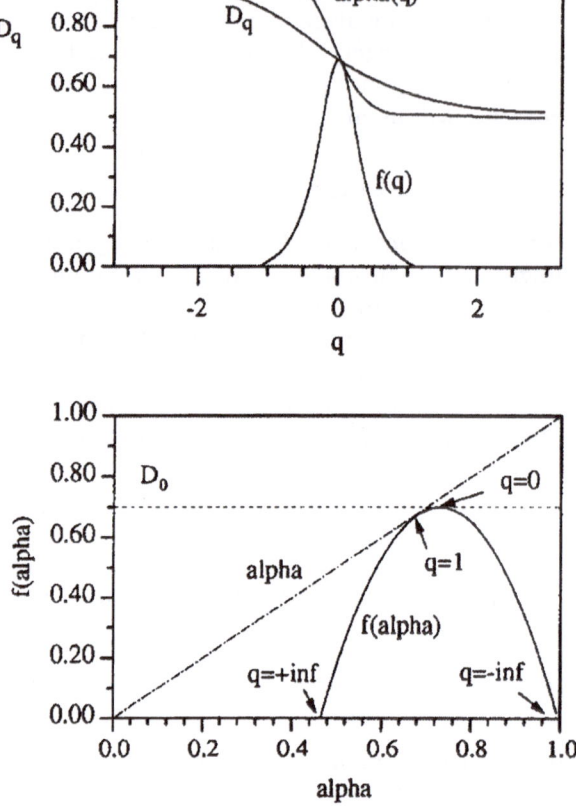

Fig. 2.6. A typical D_q curve for a multifractal and the parametric curves $f(q)$ and $\alpha(q)$ (top) and the $f(\alpha) - \alpha$ curve and its geometric relation with D_q (bottom) (from Kruel (1992))

curvature of D_q. The introduction to the latter spectrum, however, is also not to be repeated here (see [Gol96] instead) but it should be mentioned that the $f(\alpha) - \alpha$ curve has a physical interpretation attached which the spectrum of generalised dimensions lacks. The meaning of the former curve will also become apparent in the sections were it is obtained from earthquake data later on.

Practical Determination of Multifractal Spectra. Problems and errors associated with the determination of D_q multiply in comparison to the difficulties associated with the estimation of a single fractal dimension.

Equations (2.19) and (2.20) may be directly exploited to calculate the generalised dimensions simultaneously (i.e. by going through the procedure of covering only once) by computer ([Sar92, LT89, PS87]). Although the method of generalised sphere counting as outlined in [Gol96] is preferable especially for small data sets and high embedding dimensions (e.g. > 3), a method based directly on box-counting may be used for large data sets in low embedding dimension (such as one-dimensional dusts of earthquake event times or epicentre distributions, for example). The latter approach is outlined in appendix A. Another more complex idea is based on nearest neighbour information ([FPH90, PBJD79, vdWS88]); although the latter method has not been employed to obtain the results given in this work, its application to several (synthetic) data sets yielded results that make the method seem worth further investigation in the future. As has been said earlier, the multifractal spectra were obtained from the spectra of generalised dimensions in each case. This is the usual way followed in the literature in spite of a method published by Chhabra (1989) which permits the direct determination of $f(\alpha)$. The latter method should definitively be compared to the above mentioned ones in future studies.

A problem especially prevalent when estimating multifractal spectra is the determination of the scaling region—choosing a slightly different lower and upper scaling limit has profound influence on the obtained slopes and probably accounts for the differences found in published results for similar or even the same data sets. A non-linear optimisation approach for the automatic simultaneous detection of scaling region and exponent has been described in [Gol96] and used throughout this work. The issue of discrimination between spuriously multifractal data and truly multifractal data is also addressed in [Gol96].

2.2.7 Deviations from Ideal Fractals

Several deviations from ideal fractal behaviour exist as has been indicated earlier already. Usually, these "anomalies" are ignored in published fractal analyses not only of earthquake data. The latter is unfortunate not only because results may be seriously altered and thus rendered meaningless, but it also ignores additional information which might in fact be more significant than the scaling exponents themselves. A brief overview of such deviations has been given by Goltz (1996), a more detailed discussion is presented by Hastings and Sugihara (1993).

Here is nevertheless a concise summary of possible deviations because some of these features will be used extensively in the chapters to come.

As has been outlined in the course of introduction of the different important fractal dimensions, one important "deviation" is the non-uniformity of fractals. The latter is only recognised during a multifractal analysis which has been described above and elsewhere. Deviations which may occur independently of mono- or multifractality include:

– Limited scaling region
    In theory, a fractal scales from infinitely small scales up to the overall size of the object. In practice, the size of the data set is limited, leading to upper and lower scaling limits. During the determination of the scaling exponent, the sufficient size of the scaling region must be ensured. Usually,

a range varying over a factor of 10 is considered sufficient to believe in the fractal structure (also e.g. a circle produces a limited linear range in the log − log plot!).

− Multiscaling

The above mentioned sufficiently large scaling region might display two (bi-fractality) or more regions of distinct slopes. This implies that different physics underlie different scales. The fragmentation of large rocks might for example produce fragments of different fractal shape or size distribution than the fragmentation of very small samples because the fragmentation process obeys different physics at large and small scales. When analysing samples of all sizes, a crossover point in the log − log plot would indicate such a transition. Averaging over such a piecewise linear curve obviously produces dubious results.

− Anisotropy

All profiles (i.e. also "time series") are self-affine, i.e. possess a different fractal dimension in perpendicular directions parallel to the x- and y-axis. This must be so because these "objects" are single-valued and the two axes usually represent very different physical quantities (e.g. time and concentration of Radon gas in groundwater). But also topographical profiles are self-affine because gravity works only in the vertical direction, while erosion and other land-forming processes work in all directions (e.g. [OM92]). More general, if D varies freely with direction, the fractal may be called (strongly) anisotrop.

In this work, sufficient size of the scaling region was assured either manually or by automatic nonlinear optimisation, all log − log curves were manually inspected for multiscaling but no such behaviour was found. The analyses of multifractal properties is a major point and detailed anisotropy analyses were carried out for several examples of seismic fields.

3. A Sample Application to Landslides

3.1 Introduction

Landslide distributions in two major areas of northern Japan, Tohoku and Hokkaido, are analysed for multifractal properties. For the latter data set, the multifractal spectrum for the spatial landslide size distribution is also determined and compared to the probability distribution. It is concluded that the fields possess definite multifractal character. This finding is supported by the known multifractality of the main triggering processes, rain and earthquakes. Further support comes from a configuration entropy analysis which is found to be a useful complimentary tool to multifractal analysis. Models leading to multifractality are briefly reviewed. Careful attention is paid to the algorithms used and to the verification of the numerical results. Some general suggestions concerning numerical methods are made.

Landslides, while historically not renowned for being as disastrous as earthquakes or tropical cyclones, have had just as dramatic an impact on property and lives. Landslides are a rapid onset natural hazard just like earthquakes but they are more widespread and thus form one of the processes responsible for the shape of the earth's surface (Scheidegger 1991). Japan lies on the border between oceanic and continental crust and therefore shows complex geological structure. Seventy percent of Japan is mountainous and the population and industry is therefore concentrated in the narrow strip between the mountains and the sea.

The approach to landslides usually concentrates on the slope failure and subsequent debris flow (e.g., [Tak91]). However, observational data under controlled conditions is difficult to obtain so that computational models are usually studied. In recent years cellular automata, lattices where the usually discrete evolution of sites depends on their immediate neighbours by local rules, have been increasingly applied to the modelling of landslides (e.g., [SD95]) and many other physical phenomena (Carlson et al. 1993). In particular the sandpile automaton by Bak et al. (1988), introduced in connection with avalanche dynamics, has made an attempt to explain the frequent occurrence of power-law statistics, i.e. fractality, in nature. The explanation is given by a supposedly universal feature, the self-organised criticality, in which systems evolve towards a critical state without any control from outside. Several automatons leading to criticality have been shown to adhere to fundamental macroscopic differential equations in their thermodynamic limit (e.g., Frisch *et al.* 1986).

Due to these relationships between thermodynamics, self-organised criticality and multifractals ([SL93]), the concept of multifractals comes naturally into play when considering landslides. Indeed Segre and Deangeli (1995) perform a multifractal analysis of the temporal behaviour of their cellular automaton for the realistic modelling of debris flow to validate it. Their line of argument is that the model must show distinctive nonlinear behaviour which can be identified by detecting intermittency, i.e. rare irregular burst of enhanced activity. Intermittent behaviour in turn requires multifractality of a signal that describes the global behaviour of the process in a suitable way (see also [PVBV93]).

All the above studies focus on individual, isolated sites. In this paper, landslides are regarded as a spatially distributed system, assuming that they might as a whole be the expression of some, possibly nonlinear, dynamic process. Basically, the idea is that, if the individual event on its local (fractal) topography and with its local geotechnical parameters shows multifractality, the whole process might have the same property on larger scales (see [Ito92] for a similar line of thought in the case of earthquakes).

While geological and topographical factors condition the location and size of landslides, the possible trigger mechanisms form a multitude of interrelated primary and secondary effects (e.g., [Cro86]). These effects include interactions between landslides on various scales which supports a dynamical system approach. The primary causes for the triggering of landslides, however, are heavy rainfalls and earthquakes, both of which have been shown to be multifractals ([SL93] and references therein, [HLS+94, HIY92, GGP90, HI91], Godano and Caruso 1995). Recently examples for the multifractality of topography itself have also been given ([SL93] and references therein). Analyses of radar reflectivity data obtained from rain fields show a strong correlation with the location of landslides (e.g., [SAK88]). These findings further strongly support the assumption of multifractality of the landslide process.

Multifractality of the spatial landslide distribution was first assumed by Fukuoka *et al.* (1994) when numerical results indicated the inequality of the first three generalised dimensions but no further analysis was carried out at that time. However, the results led to the present study in which also the spatial distribution of landslide sizes is analysed for the first time. A considerable part of this chapter is devoted to the theoretical background of fractals and multifractals and the related numerical methods as well as their verification. Finally a configuration entropy analysis ([BAL+94]) complements the multifractal analysis.

3.2 The Data

Figure 3.1 gives the geographical situation of the regions under consideration and shows the landslide distributions for Hokkaido Island and the Tohoku region.

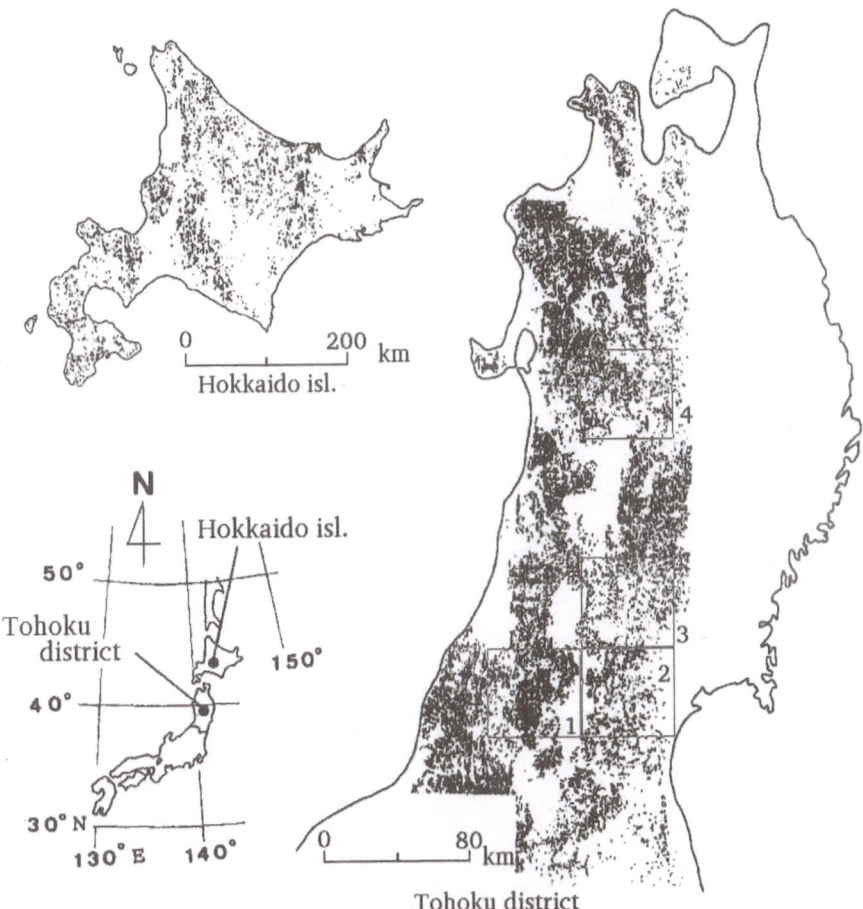

Fig. 3.1. Geographical location of regions considered and their landslide distributions. Also indicated are four subsets of Tohoku which will be referred to later (after Hiura and Fukuoka, 1994)

Both landslide distribution sets were obtained by examining aerial photographs and topographical maps and cover roughly 10 000 years of the landslide history. The location and, in the case of the Hokkaido data, the size (area) of the landslide was obtained from the center and area of an ellipsoid fitted to the amphitheatre produced by the respective landslide.

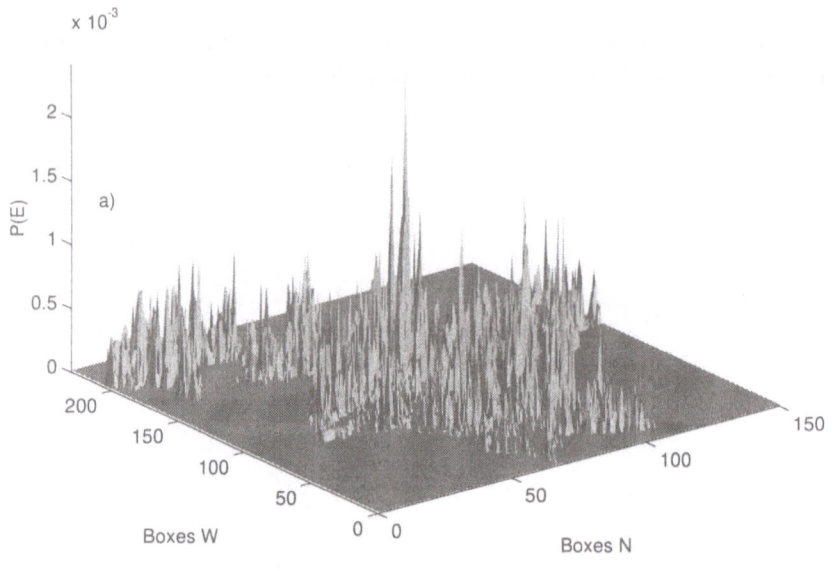

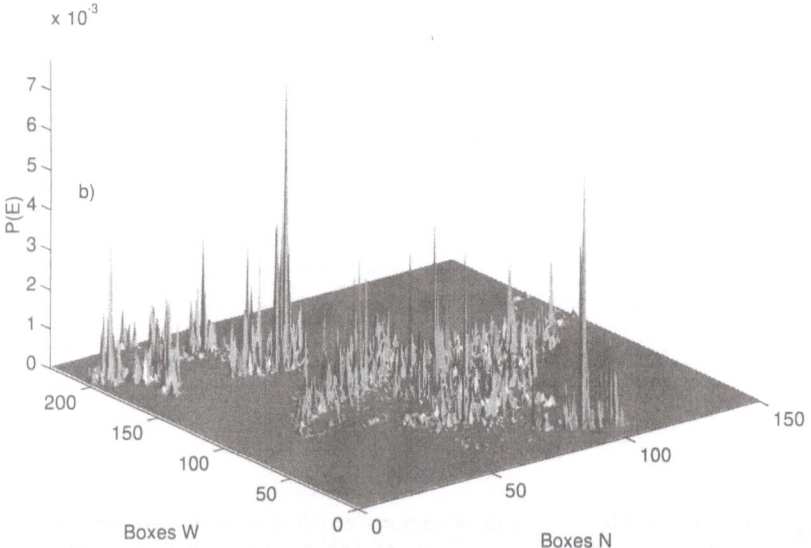

Fig. 3.2. Perspective three-dimensional views of the probability density distribution (a) and the landslide size distribution (b) in Hokkaido. The fields were generated by calculating the normalised cumulative probabilities and sizes for 3 km by 3 km cells and thus represent the respective measures at that resolution

The location error is estimated to be better than 250 m for both sets. The Tohoku data set contained 57 520 events, Hokkaido consisted of 12 842 data points where the smallest landslide had a size of 0.004023 km² and the largest had a size of 13.0017 km².

To gain an impression of the variability, or intermittency, of the landslide phenomenon, Fig. 3.2 gives three-dimensional perspective views of the probability density distribution (a) and the landslide size distribution (b) for Hokkaido. To generate the shown fields, the plane was partitioned into 3 km by 3 km cells. For each cell the cumulative normalised probability respectively the cumulative size was calculated.

In both cases, landslides are concentrated in tectonically active zones. In Hokkaido, this area is represented by the mountainous zone near the middle of the island, from north to south; in Tohoku, the densest clustering occurs in the green tuff zone along the Ohu chain and the Dewa mountains. No easy distinction can be made between the surface geologies, although Tohoku features large regions of tertiary clay layers while Hokkaido shows large regions of non-metamorphic rock.

Due to the missing elevation value in the location of the landslides, the data represents a projection onto the two-dimensional plane. However, as the results for D_0 will show, the sets posses a capacity dimension well below 2 so that it is assumed that no information is lost. This is inferred from the fact that a projection of a set into a lower dimension d will have fractal dimension $D_0 = d$ if the original set had a fractal dimension higher than d, otherwise the fractal dimension is the same (e.g., [HS93]).

3.3 Fractal Analysis

For practical purposes, a fractal is an object or set of non-integer, i.e. fractal, dimension. For natural objects the fractal dimension is an expression of statistical self-similarity or, in other words, scale invariance, which in turn can be understood to be a principle of symmetry. Thus one would expect many geophysical processes to be scaling because often no predominant mechanisms are present which could break the scaling.

The most basic definition of fractal dimension for a natural object is the one based on box-counting (e.g., [Bar88]) where the minimal covering of the set is determined at successively finer resolution: The capacity dimension is defined to be

$$D_0 = \lim_{r \to 0} \frac{\log N(r)}{\log(1/r)}$$

where $N(r)$ is the number of non-empty boxes of size r. The information dimension D_1 and correlation dimension D_2 are two more fractal dimensions which are frequently encountered in the literature and which have specific interpretations attached (e.g., [Fed88]). D_2, respectively an algorithm for its determination, the Grassberger-Procaccia algorithm ([GP83]), often also called sphere-counting, was shown to be more robust a measure of fractality because it weighs more heavily the denser regions of a set whereas the box-counting method weighs equally all non-empty boxes regardless of the number of points contained.

The above three dimensions are a small subset of infinitely many generalised dimensions (Halsey *et al.* 1986) of the form

$$D_q = \frac{1}{q-1} \lim_{r \to 0} \frac{\log(\sum_i \{P_i(r)\}^q)}{\log r} \tag{3.1}$$

where $P_i(r) = N_i(r)/N$ is the probability or density (or, more general, any measure or "mass") in the ith box. For $q = 0$, boxes are weighted only according to whether they contain points of the set or not, leading to the capacity dimension, for $q = 1$ the information dimension is obtained and the correlation dimension follows for $q = 2$. As will be seen below, for a simple self-similar fractal (a homogeneous fractal or monofractal), $D_q = const.$ for q being a real value in the range $-\infty \leq q \leq \infty$. $D_q > D_{q+\Delta q}$ is characteristic of a multifractal and will be the case of interest in this paper.

The formulation in Eq. (3.1) has been coined "strange attractor notation" because it was first applied to probability distributions on strange attractors in the phase space of nonlinear dynamic systems. Here, an alternative formalism, the multifractal $f(\alpha)$–α spectrum, will be introduced later on because the generalised dimensions offer no direct physical interpretation.

In practice, D_q is determined from the slope of a linear region (the scaling region) in a $\log(\sum_i \{P_i(r)\}^q)$ versus $\log r$ plot. Ideally, the scaling region extends from the minimum distance of any pair of points in the set to the overall size of the set. With experimental data, the scaling region is often much smaller, leading to uncertainty in the determination of the scaling exponent. An additional possible deviation is that two or more piecewise linear regions might be present instead of only one, yielding several different scaling exponents at different scales. The latter case is termed multi-scaling as opposed to multifractal (although, unfortunately, some confusion exists in the literature about these terms).

Another, usually not considered, deviation from ideal self-similar behaviour is self-affinity or, more general, free variation of the fractal dimension with direction (anisotropy of the fractal dimension). While time series and profiles are necessarily self-affine (e.g., [Fed88]), all other fractals occurring in nature might also have this property. Examples are rain fields ([SL93] and references therein) and other atmospheric phenomena due to the stratification of the atmosphere and intrinsically directional processes like machining and wear of material surfaces ([Rus94]). Anisotropy could be introduced into the landslide fields when landslides would happen along several predominant superimposed ridges for example. Self-affinity may be artificially introduced by the unequal rescaling of the axes of finite data sets, too. Such anisotropy does not show up during the standard methods of fractal analysis. Instead, depending on the algorithm used, some average or extremal result will be found. A test for directionality of the scaling exponents should therefore be included in every fractal analysis. A simple way to achieve this is the analysis of several profiles in different directions.

Box-counting is not affected by boundary effects, while the effect on sphere-counting depends on the density of points near the boundaries. If the regions near the boundaries are densely populated, more centres of disks will lie close to the boundaries and the sphere counts will consequently be underestimated because part of the disk falls outside the limits of the data set. A comparison between the value of D_0 obtained by box-counting and by sphere-counting might be useful.

Figure 3.3 shows the log–log plots for the landslide distributions in Hokkaido (circles) and Tohoku (crosses). Linear regression over the range of 2.73 km to 174.57 km gives D_0 = 1.57 ± 0.02 for Hokkaido and D_0 = 1.65 ± 0.03 over a range of 0.87 km to 110.80 km for Tohoku (errors given are one standard deviation of the slope). The scaling regions extend

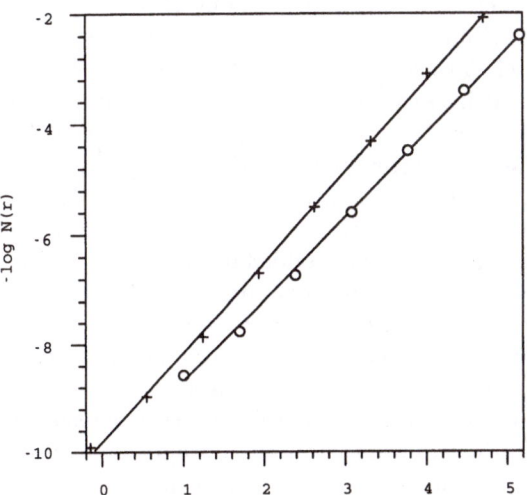

Fig. 3.3. Plot of $\log(1/N(r))$ versus $\log r$ for the landslide sets of Hokkaido (circles) and Tohoku (crosses). The solid lines show the least square fits the slope of which gives D_0

over ranges which vary by at least a factor of about 60, which can be considered sufficient to believe in the fractal structure of the data. No multiscaling behaviour is apparent which shows that there are no different superimposed physical processes acting at different scales. Furthermore no directionality of D_0 was found. The latter two findings might be attributed to the high complexity of the phenomenon due to the underlying factors which possess no directional or predominant features at different scales.

3.4 Multifractal Analysis

Above it was shown that the data possesses a fractal structure. However, there are different degrees of clustering in the landslide distribution meaning that the distribution differs from the neighbourhood of one landslide to the neighbourhood of another. In a homogeneous fractal all neighbourhoods are similar. Differences between neighbourhoods can be described by the concept of multifractality if the irregularities have statistical similarity over a sufficient range of scales.

A concrete idea of a multifractal can be obtained from the distribution of mineral in the earth's crust in a certain region S: Imagining a measure μ which denotes the amount of mineral contained in every subregion, it is certainly expected that this measure is not regular: Dividing the original region into two halves S_1 and S_2, one expects $\mu(S_1)$ and $\mu(S_2)$ to be different. Dividing S_1 again, S_{11} and S_{12} again contain different amounts of mineral. This unequalness is valid down to very small rock samples. μ is a measure which is irregular at all scales. The irregularity of mineral distribution furthermore is statistically the same at all scales, it is a self-similar measure or multifractal as was shown by the mineralogist De Wijs as early as about 1950.

De Wijs (1953) found the distribution of mineral in rock to be well approximated by the following model: A rock contains a certain mineral of total mass (measure) μ. If the rock is cut into two, one half contains a fraction $p\mu$ of the mineral, the other half contains $(1 - p)\mu$. Splitting the left half of the rock into equal volumes again, the left quarter contains $p^2\mu$, the right quarter contains $p(1 - p)\mu$. At every stage of the construction, every part is divided into two equal parts and the mineral is redistributed unevenly according to the same rule. Thus, at every step, the mineral is divided in the ratio $p : 1-p$. The right two quarters of the second stage hence contain mineral of masses $(1 - p)p\mu$ and $(1 - p)^2\mu$. This model, which for $p \neq 1/2$ (as observed experimentally) leads to a deterministic multifractal on a non-fractal support, is used later on to produce artificial data sets. If the above process is imagined to take place in the plane and the total mass of mineral is unity, the result is a probability measure and the amount of mineral in every fragment may be thought to represent the number of landslides in every subregion.

In the following the concept of multifractality is quantitatively introduced in a simplified way. The introduction mostly follows Feder (1988) and Emmerson and Roberts (1994), i.e. the multifractal spectrum ("f–α-formalism") is used instead of the related generalised dimensions ("D_q-formalism").

The Lipschitz-Hölder Exponent α

Clearly some measure is needed to quantify the different degrees and kinds of clustering around every landslide. The Lipschitz-Hölder exponent α attempts this by describing the scaling properties of the neighbourhoods. If there are

$N_j(r)$ landslides within the neighbourhood of radius r around the jth land-slide, the fraction $\nu_j(r) = N_j(r)/N$ of all landslides is contained in this disk. If $\nu_j(r)$ scales with r like $\nu_j(r) \propto r^\alpha$, i.e. the neighbourhood is a fractal, then, for $r \to 0$, the scaling exponent α is a local property peculiar to the jth landslide. Hence the name local fractal dimension for α. As α is actually the correlation dimension D_2 (cf. [GP83]), it is obvious that α must be estimated from a significant scaling region somewhere between the minimum distance in the set and its overall extent rather than from "$r \to 0$": insisting on "$r \to 0$" renders all finite sets to have dimension 0 (which is true as the self-similarity breaks down below the minimum resolution).

For a homogeneous fractal, α is the dimension of the set, which is what the Grassberger-Procaccia algorithm and all other pointwise dimension esti-mators are based on. Examples of homogeneous sets are equidistantly placed points on a line ($\nu_j(r) \propto r^1$, 1 being indeed the dimension of the line), points placed on a regular grid in the plane ($\nu_j(r) \propto r^2$, 2 being the dimension of the Euclidean plane) and the classical two-thirds Cantor set ($\nu_j(r) \propto r^{\log 2/\log 3}$). Examples for non-uniform distributions in one dimension can be easily con-structed by placing points at $x_j = \pm|j|^p$: starting at point $x_0 = 0$, it is $N_0(r) \propto r^{1/p}$, that is $\alpha = 1/p$. For $p = 2$ the points are quadratically in-creasingly farther apart with increasing distance from the origin. α associated with x_0 is 0.5. For $p = 1/2$, on the other hand, points are spaced closer and closer together the farther away from the origin. Here α is 2, which is twice the dimension of the embedding space! Thus, roughly, for a non-uniform dis-tribution, a low value of α corresponds to a dense cluster in a less populated surrounding and a high value means that a sparse area is surrounded by a vicinity of denser clusters. Hence also the names crowding index, singularity strength or singularity index for α.

The Multifractal $f(\alpha)$ Spectrum

As every landslide has a value of α associated with it, one now has just another set of numbers if the distribution is non-uniform. One approach to describe this set more conveniently is the multifractal spectrum in which the fractal dimension f of all landslides with a common α is calculated. The $f(\alpha)$ curve is the multifractal spectrum. Thus the original data is decom-posed into several subsets of landslides with common distributions in their immediate neighbourhoods. This is why a multifractal may be called a union of intertwined fractals of different dimension. It should be noted, however, that multifractality is a property of a measure, not of a set itself. The De Wijs multifractal mentioned earlier is defined on an Euclidean support for example. In the uniform case α is independent of the location of the landslide and thus $f = \alpha = D_0$ and the $f(\alpha)$ curve collapses into a single point. The D_q curve becomes a straight line through D_0 parallel to the abscissa. Only then a single fractal dimension suffices to describe the fractal.

Figure 3.4 shows the theoretical $f(\alpha)$–α curve for the De Wijs fractal in two dimensions with $p = 0.25$. The curve can be interpreted as follows: The

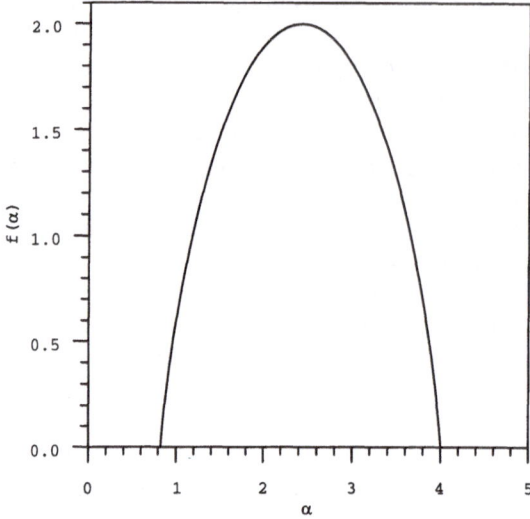

Fig. 3.4. Theoretical multifractal spectrum of the two-dimensional De Wijs fractal with $p = 0.25$

spectrum, which looks like an upside down parabola, peaks at $f(\alpha_0) = D_0$ and stretches from α_{min} to α_{max}. Recalling that the fractal dimension tells how densely a set occupies its embedding space, it becomes clear that most landslides (assuming for now that the model of De Wijs does not redistribute mineral but landslide events) contribute to the peak as this is the highest fractal dimension occurring in the union. Thus, if $\alpha_0 > 2$, most landslides are surrounded by denser clusters, if $\alpha_0 < 2$, most landslides are situated among sparser regions. For $p = 0.25$, most events are therefore surrounded by denser clusters. The large α side is generated by landslides in sparsely populated areas. The lower value of f, i.e. the lower fractal dimension of the set consisting of landslides in low density regions, shows that there are fewer of them. The low α side describes densely clustered landslides and similarly there are fewer of them. In summary, the range α_{min} to α_{max} quantifies the non-uniformity of the fractal while $f(\alpha)$ tells how frequently events with scaling exponent α occur.

Mass Exponents

Unfortunately the computational determination of the multifractal spectrum is much more difficult than the often already ambiguous determination of one single fractal dimension. This becomes especially evident when the (multi)fractal properties of different sets are to be compared and conclusions as to the underlying dynamics or the universality of the phenomenon are to be made.

Here the description by Feder (1988) is adapted in which the embedding plane is partitioned into squares of size δ first and only non-empty boxes are kept. If $N_i(\delta)$ is the number of landslides within the ith box, the mass $\mu_i = N_i/N$, where N is the total number of landslides, is assigned to the ith box. The mass values on their support of non-empty boxes constitute the measure (here the probability measure or natural measure in strange attractor terminology) which is analysed for multifractality, not the set itself.

Having obtained the measure, which here constitutes a geophysical field, one calculates for various exponents q

$$C(q,\delta) = \sum_i \mu_i^q \qquad (3.2)$$

and attempts to determine the mass exponent $\tau(q)$ from

$$C(q,\delta) \propto \delta^{-\tau(q)} . \qquad (3.3)$$

$\tau(q)$ is determined by plotting $\log C$ versus $\log \delta$. Due to its usage in thermodynamics, $C(q,\delta)$ is frequently called partition function. The effect of the normalisation $\sum_i \mu_i = 1$ is that $\tau(1) = 0$ and $\tau(0) = D_0$ because for $q = 0$, C is just the count of boxes needed to cover the set. For measures other than the natural one, normalisation to a probability measure is also essential because otherwise the possibly existing scaling will not be discernible.

For increasingly large positive q, large values of μ_i contribute increasingly more to the qth statistical moment $\sum_i \mu_i^q$ while small values get weighted less and less. Thus more emphasis is put on the densely clustered landslides respectively the larger events. On the other hand, for increasingly negative values of q, the sparse regions (or small events) dominate the moments. Hence the fractal dimension is determined for separate subsets consisting of points associated with different magnitudes of μ_i. The latter formulation leads to a simple test for multifractality (cf. [HS93]): One may check whether e.g. the correlation dimension changes when successively thresholding the data.

Using statistical moments, one can describe the multifractal by recovering information about the different regions with different scaling exponents by examining the variation of the mass exponent $\tau(q)$ with q. α and f can then finally be obtained parametrically from the relations

$$\alpha(q) = -d\tau/dq$$

and

$$f(q) = q\alpha(q) + \tau$$

([Fed88]) or, explicitly and computationally more stable, from

$$f(\alpha) = \min_q \{q\alpha + \tau(q)\} .$$

Having in mind the often encountered better robustness of sphere-counting as compared to box-counting, a slightly different approach is followed here in the determination of $\tau(q)$. Namely, the approach is based on the correlation dimension, leading to the method of generalised sphere counting as proposed by Pawelzik and Schuster (1987): Recalling the context of Eqs. (3.2) and (3.3), a disk of radius r around the jth landslide contains $N_j(r)$ landslides and thus the mass $\mu_j = \nu_j = N_j/N$. A fraction $1/N_j$ of the disks contents is needed to cover the jth landslide itself. Thus the jth landslide contributes $(1/N_j)\nu_j^q$ to the sum in Eq. (2). One may thus calculate

$$C(q,r) = \frac{1}{N}\sum_{j=1}^{N}\nu_j^{q-1}.$$

Then, plotting $\log(C^{1/(q-1)})$ versus $\log r$, the linear region will yield $D_q = \tau/(1-q)$. Thus D_q is estimated rather than $\tau(q)$.

The method for multifractal analysis employed here is, like many other methods (e.g., Borgani et $al.$ 1993), dependent on the convergence of statistical moments. The divergence, i.e. non-existence, of a moment does not necessarily require an infinite singularity but may be caused by a power-type long time tail behaviour as well: In the case of $1/f$ noise for example, the variance is infinite. Indeed also the Gutenberg-Richter power-law of earthquake magnitude versus frequency predicts the divergence of statistical moments ([HLS$^+$94]). Even if they exist, however, statistical moments get less robust with increasing order (e.g., [P$^+$92]). Therefore, careful attention was paid to the convergence of the statistical moments involved. Namely, it was assured that the moments were finite, showed convergence as the number of data points was increased and that they were consistent for different data sets obtained from the same distribution.

Determining the region and the exponent of scaling The log–log plots consist of a linear region sandwiched between two horizontal regimes. The horizontal regime for low r results from the finite resolution of the data below which the number of points stays constant for decreasing r. The horizontal high r region results from the finite size of the data where no additional points are found for increasing r. As the linear range is unknown beforehand and varies from data set to data set and with q, it is not advisable to perform a blind linear regression over a fixed range. A reliable and automatic way which also determines the scaling region is instead desirable.

A possible solution ([ER94]) is to fit the data by a transcendental curve of the form

$$\eta = \bar{\xi} + D\frac{1}{2}\{\zeta_1\log[2\cosh((\xi-\xi_1)/\zeta_1)] - \zeta_2\log[2\cosh((\xi-\xi_2)/\zeta_2)]\}$$

where $\xi = \log_{10}r$, $\eta = \log_{10}(1/C)$ were substituted, $\bar{\xi} = \frac{1}{2}(\eta_{+\infty} + \eta_{-\infty})$ and $D = (\eta_{+\infty} - \eta_{-\infty})/(\xi_2 - \xi_1)$. These functions have the following properties: The graph is horizontal and asymptotically approaches $\eta = \eta_{-\infty}$ for $\xi \ll \xi_1$; for $\xi \gg \xi_2$ it is also horizontal and approaches $\eta = \eta_{+\infty}$. In the range $\xi_1 \ll \xi \ll \xi_2$ it is asymptotically a straight line of slope D. ζ_1 and ζ_2 describe the widths in η of the exponential transitions between the three domains. Thus, after minimisation of the root-mean-square distance between the transcendental curve and the data, D gives the fractal dimension and ξ_1 and ξ_2 give an estimate of the lower and upper limits of the scaling region. Confidence in the result for D may then be obtained from the range of ξ_1 and ξ_2 which should span at least a factor of 10.

As the curve depends nonlinearly on the four parameters ξ_1, ξ_2, ζ_1 and ζ_2, nonlinear optimisation must be utilised. Several different approaches to nonlinear optimisation exist (e.g., [P+92]) and a comparison mainly showed different robustness with respect to the accuracy of initial guesses. The Marquardt-Levenberg method ([P+92]) was found to be reasonably reliable even under slightly disturbed conditions (see below). While large absolute values of q still required manual checking, results for the range $0 \leq q \leq 5$ were reliable in all cases when the optimisation was modified as to avoid degenerate curve fits of the form $\xi_1 + \zeta_1 \geq \xi_2 - \zeta_2$.

Figure 3.5 shows the $\log(C(q,r)^{1/(q-1)})$ versus $\log r$ plots together with the automatically determined best nonlinear fits for $q = 15, 0, -15$ (from top to bottom) for the Tohoku landslide distribution. The raw data is drawn in solid, the fitted curves are dashed. The automatically determined scaling regions were 0.22 km to 105.95 km ($q = 0$), 0.16 km to 64.90 km ($q = 15$) and 4.76 km to 129.28 km ($q = -15$). Note that the optimisation algorithm

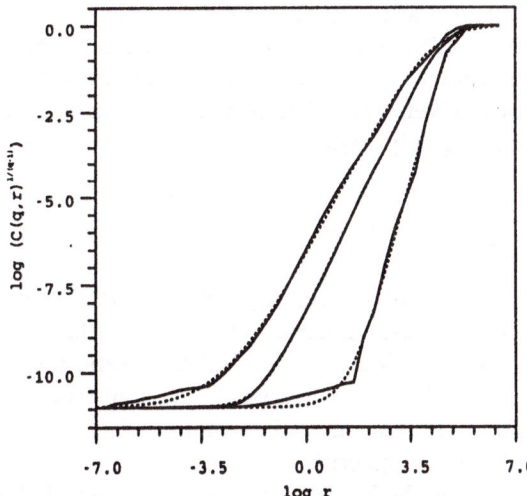

Fig. 3.5. $\log(C(q,r)^{1/(q-1)})$ versus $\log r$ plots (solid) together with the automatically determined best nonlinear fits (dashed) for $q = 15, 0, -15$ from top to bottom for the Tohoku landslide distribution

is capable of dealing with the anomalous behaviour at low r for $q = -15$ and that the results agree with an optimal manual choice to such an extent that they can be used safely to gain confidence in the fractality of the data.

3.5 Analysis of Synthetic Data

While the problem of accuracy of fractal dimension estimates from finite data sets and the definition of confidence limits has received much attention ([BMPV93, Gra88, RY90, HH94]), no readily applicable rules are available so far. The standard deviation of regression parameters tells nothing about possible methodical errors for example (cf. [GMM98]). Here a direct comparison between the results obtained from experimental data and from mono- as well as multifractal synthetic data with the same overall characteristics is carried out therefore. Performing the same numerical analysis on these data sets, one may estimate the range of results obtainable from these fractal models. All sets are generated to have the same number of data points and to occupy the same space in the plane to exclude possible differences due to size and spatial extent.

Artificial Monofractals

Numerous methods exist to generate data sets of a specific fractal dimension of the support (e.g., [PS88, HS93, Rus94]). Most simple are deterministic sets like the Sierpinski triangle or the Cantor set. The dimension of all deterministic fractals can be changed by modifying the generator. A general approach to the modelling of arbitrary fractal objects which also yields more realistic sets is the one by affine transformations. Random fractals are more realistic alone by the fact that most fractals in nature are not deterministic.

Here Iterated Function Systems (IFS; [Bar88], [GW88]), each consisting of N contractive affine transformations with associated probabilities, were used to produce monofractals of a given D_0. Monofractals as opposed to multifractals were obtained by utilizing only angle-preserving transformations so that

$$\sum_{i=1}^{N} |s_i|^{D_0} = 1 \qquad (3.4)$$

(e.g., Feder 1988), where $s_i = const.$ are the scaling factors of the maps, could be exploited to get D_0 as a convenient function of s_i only. So far, Eq. (3.4) has been used to estimate D_0 of self-similar deterministic sets, i.e. in the inverse problem. The approach delivers an overwhelming multitude of optically different fractals not only with the same dimension of the support but also with equally narrow multifractal spectra.

Thus the weakness not only of a single fractal dimension but also of the multifractal spectrum in distinguishing optically distinct fractals is demonstrated. On the other hand, all these fractals certainly belong to the same "universality class" insofar as they are generated by the same principle. In that sense the multifractal spectrum allows to classify them. The problem of the seemingly contradictory goals of distinguishing fractals and classifying them by multifractal spectra is addressed again later on. The familiar Sierpinski triangle can be obtained from an IFS with $N = 3$ and $s = 0.5$. Equation (3.4) then correctly gives $D_0 = \log 3/\log 2$. Sets of given spatial extent were obtained here by assuring that the attractor was larger than needed and then a zoom window of desired size was filled with the needed number of points. This agrees with the fact that experimental observations usually also represent a subset limited by the extents of the area of observation.

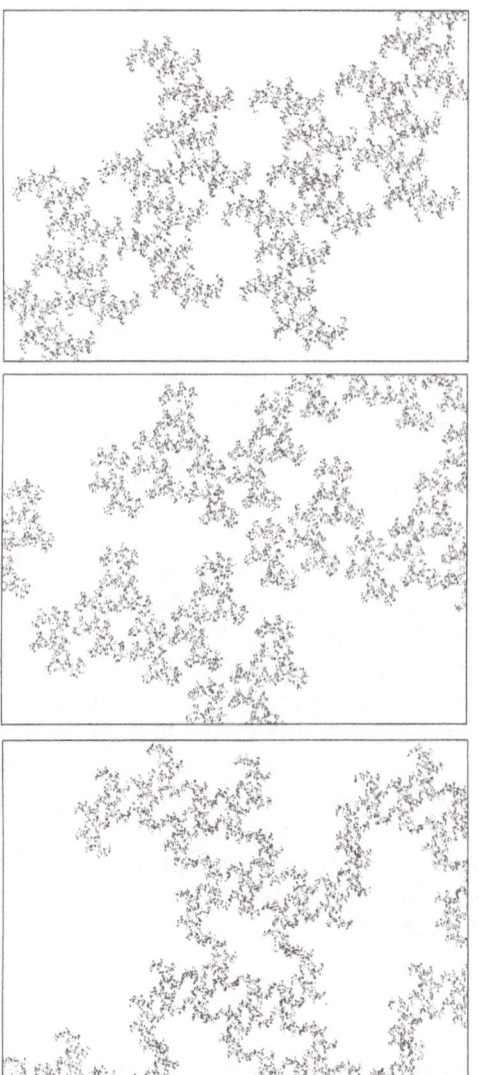

Fig. 3.6. Three artificial monofractals with the same overall characteristics as the landslide distribution in Hokkaido. Note that these sets are perfectly equivalent to each other and the Hokkaido data as for the fractal properties of the support and that a multifractal analysis cannot distinguish the artificial sets

Figure 3.6 gives three examples of IFS-monofractals generated to fit the overall characteristics of the Hokkaido data set ($D_0 = 1.57 \pm 0.02$). They will be referenced as sets F1, F2 and F3 (from top to bottom) further on.

Note that these sets are perfectly equivalent to the Hokkaido data as for the fractal properties of the support. Furthermore one would expect equivalence of the multifractal spectra for the artificial sets which is confirmed in Fig. 3.7, where the numerically determined $f(\alpha)$–α curves are shown. One

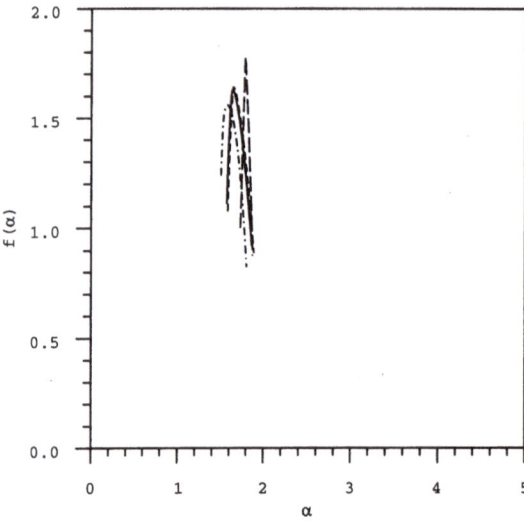

Fig. 3.7. Multifractal spectra of the artificial monofractals shown in Fig. 3.6 (solid: F1, short dash: F2, dash dot: F3). Also shown is the spectrum for a deterministic Sierpinski triangle with $D_0 = 1.8$ (long dash)

notices good agreement of the location of the maxima with the expected value $f = \alpha = D_0 \approx 1.57$ and that the spectra are barely distinguishable. For comparison, the spectrum of a deterministic Sierpinski triangle with $D_0 = 1.8$ was determined and the resulting maximum is also found to be very close to the theoretical value. The finite range in α and the observed variation in the location of the maximum is due to the limited amount of data points. It is interesting to note that $\Delta\alpha$ is narrower for the deterministic construction than for the random fractals, implying that the accuracy of the determined spectrum does not depend on the number of data points alone. In fact more points are required to determine a higher fractal dimension with the same accuracy as a smaller one ([HH94] and references therein). Overall, $\Delta\alpha$ is found to be smaller than 0.5.

Artificial Multifractals

IFSs usually produce multifractal-like measures and may be used to produce given multifractal measures (e.g., [EM92]) but here a direct deterministic approach is preferred. Namely the De Wijs model for the distribution of mineral in rock is used. In physics, the model is known as binomial multifractal or binary multiplicative fractal and is the simplest multiplicatively generated measure. The process, which fragments a set (e.g. the rock) into smaller and smaller units according to a fixed rule (halve it at each step), and simultaneously fragments the measure of the units (e.g. the mineral) according to another one, is also called a cascade process. The idea of cascades has found application in the modelling of nonlinear dynamical systems like fully developed turbulence where energy is passed on to smaller and smaller eddies until it is finally dissipated. An extensive treatment of the multifractal aspects of the binomial multifractal may be found in Feder (1988).

An extension of the binomial De Wijs construction leads to multinomial multiplicative processes and finally to random multiplicative cascades where the multipliers for the mass redistribution are the outcome of some probabilistic process. While random cascades produce more realistic measures, physical models often remain elusive ([EM92]). The most realistic self-similar measures are probably produced by nonlinear stochastic processes as expressed by stochastic differential equations (Provenzale *et al.* 1993).

For numerical purposes, however, the binomial measure is most suitable as, due to the simple deterministic construction, the $f(\alpha) - \alpha$ spectrum for the case of redistribution of mass on a line is given by ([Tak90])

$$\alpha_q = -\frac{p^q \log_2 p + (1-p)^q \log_2(1-p)}{p^q + (1-p)^q} \, ,$$
$$f(\alpha_q) = q\alpha_q + \log_2(p^q + (1-p)^q)) \, .$$

For $p = 1/2$, the distribution is uniform, i.e. the construction yields a monofractal.

One can easily generate a multifractal distribution of points in the plane: In Fig. 3.8, a set of 12 842 points is shown where the x and y co-ordinates were chosen independently and at random from the one-dimensional distribution ($p = 0.25$). The right-hand sides of the above equations have to be doubled in this case.

A peculiarity of the binomial measure is that the extremal values α_{min} (highly clustered immediate vicinity) and α_{max} (sparsely populated immediate vicinity) are typically found in the left-most and right-most subintervals respectively. In general the minimal and maximal Lipschitz-Hölder exponents can lie anywhere in the support.

In Fig. 3.9, a comparison between the theoretical (solid) and two numerically determined multifractal spectra (dashed) is shown. The two numerically determined spectra stem from the set shown in Fig. 3.8 and another random subset of the same size. The scaling exponent α can be seen to vary between about 0.8 for highly clustered points and about 3.5 to 4.4 for least clustered points. As is shown by their low fractal dimension f, these extremal clusters occupy very little space of the fractal; they occur very rarely. Much more common are regions characterised by an α of about 2–3, showing a fractal di-

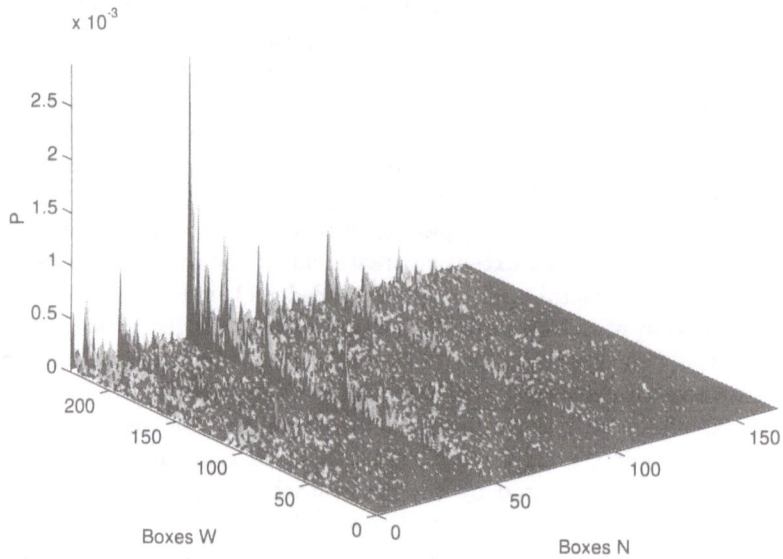

Fig. 3.8. Multifractal point distribution according to the De Wijs model for $p = 0.25$. The fractal has the same number of data points and the same spatial extent as the landslide distribution on Hokkaido

mension of ≈ 1.8 to 2. This range of α is characteristic of clusters interspersed with denser clusters.

Comparing the theoretical with the numerically obtained curves, it becomes obvious that, at least at 12842 data points, the experimental result has to be interpreted carefully: It is quite accurate for small values of α, while f becomes increasingly uncertain for the right half of the curve. α_{max} also has an error of about 10%. The position of the peak is very close to the expected D_0 of the support (2.0) in f and also reasonably placed in α. The larger error for higher values of α is due to the fact that these estimates are based on few sparsely populated regions, thus leading to weak statistics. The left side of the curve is based on strongly clustered regions, leading to more reliable statistics because more data points are available.

Similar errors are expected for the landslide distribution data even though the Tohoku data consists of roughly four times the number of data points. This is because of the previous finding that the multifractal properties of random fractals seem to be more difficult to determine than the ones of deterministic constructions. Note, however, the definite distinction between monofractals and multifractals when comparing Figs. 3.7 and 3.9. The comparison of results obtained from experimental data with those from synthetic monofractal data thus permits to confirm or reject multifractality while analysis of artificial multifractal data allows the rough estimation of the error in f and α.

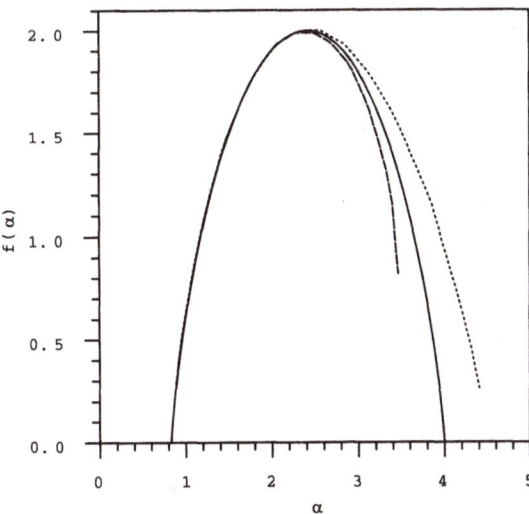

Fig. 3.9. Multifractal spectra for the data shown in Fig. 3.8 and another random set of the same size (dashed). Also shown is the theoretical spectrum (solid)

3.6 Multifractal Results for Hokkaido and Tohoku

The numerically determined multifractal spectra for the landslide data are given in Fig. 3.10. All spectra can be said to be typical in the sense that they follow the shape of an upside down parabola (this is not always so) and lean to one side ([EM92]).

Considering the curves for the probability measures first, one notices very close agreement between the values of $f_{max} = f(\alpha_0)$ and the results of the box-counting analysis performed earlier (1.58 vs. 1.57 for Hokkaido and 1.64 vs. 1.65 for Tohoku). Box- and sphere-counting thus yield the same results for D_0 here, leading to great confidence into the obtained value. The surprisingly accurate positioning of $f(\alpha_0)$ in turn results in high confidence into the respective values of α_0, especially when taking the results shown in Fig. 3.9 into account. α_0 is about 1.9 for Hokkaido and about 1.8 for Tohoku. Both values are smaller than the dimension of the embedding space, thus most landslides in Hokkaido and Tohoku are surrounded by sparser clusters. The effect is more pronounced for Tohoku though.

α_{min} is estimated to be about 1.1 for both cases and describes the areas of most intensive clustering (α_{min} corresponds to D_∞). Thus, both regions concerned possess the same degree of most intensive clustering. Interpretation of α_{max}, corresponding to $D_{-\infty}$ and characterising the least populated vicinities, is more difficult due to the large error expected. Assuming an error of also about 10%, the values of α_{max} are essentially equal. Note, however, the different overall shapes of the high-α sides, indicating that the landslide distribution in Hokkaido generally contains more regions of sparse clustering than Tohoku. The probability measure in Hokkaido can thus be regarded to be more inhomogeneous than the one in Tohoku. The landslide process in

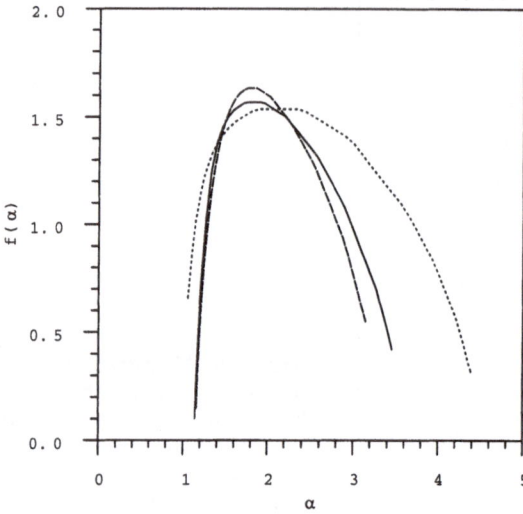

Fig.   3.10.   Multifractal spectra for the landslide data: Probability measures for Hokkaido (solid) and Tohoku (long dashed) and landslide size measure for Hokkaido (short dashed)

Hokkaido shows stronger spatial intermittency as for the occurrence probability. Nevertheless assuming the range from α_{min} to α_{max} to be roughly 2.4 for both distributions and comparing this with the value obtained for the artificial monofractals, one may be confident that the distributions really are multifractals. Comparing the results in Fig. 3.10 with the $f(\alpha)$-α curve in Fig. 3.4, it can be seen that the landslide probability distributions are less inhomogeneous than the artificial binomial distribution. This agrees with the optical impression obtained when comparing the three-dimensional visualisations in Figs. 3.2a and 3.8.

Finally, the multifractal spectrum for the spatial distribution of landslide sizes in Hokkaido can be seen to be much more inhomogeneous than the probability distributions and even the multiplicative binary fractal: α varies over a range of at least 3.5. This is confirmed when examining Figs. 3.2a,b and 3.8. The finding might be expected because the size distribution of landslides follows a power-law over a wide range of sizes ([FHG94], see also [HIY92] for the case of earthquakes).

The scaling regions used for the determination of the slopes varied between the values shown in Fig. 3.3 down to a factor of 20 at worst for the Hokkaido size distribution.

3.7 Configuration Entropy

Beghdadi *et al.* (1994) found that multifractal spectra determined by the sandbox method (e.g., [Vic92]) could not distinguish data sets with notably different optical properties. This is not surprising, as even multifractal spectra do not yield a complete but only an optimal description in the thermodynamic sense (e.g., [McC94]). Beghdadi *et al.* (1994), hereafter referred to as paper 1, applied the idea of configuration entropy (e.g., [PS88]) to their data and found it less ambiguous and less sensitive to the finite size of data than the sandbox method.

In this entropy analysis, the landslide data is regarded to be a point process and to constitute a binary image with pixels either on (active, no event) or off (inactive, representing a landslide event). Mostly following paper 1, the data is analysed as follows.

Cells of size r are centred on a random subset of inactive pixels and for each cell the number of active pixels is determined. Then $N_k(r)$, the number of cells containing k active pixels, is calculated. The probability for a cell to contain k active pixels is then $P_k(r) = N_k(r)/N(r)$ where $N(r)$ is the total number of cells of size r. The configuration entropy is then given by

$$H(r) = -\sum_{k=0}^{r^2} p_k(r) \ln p_k(r)$$

and is a measure for the uncertainty in the realisation of a certain state for a cell of size r. $H(r)$ describes probabilities distributed over boxes of fixed size of the partitioned space the set occupies. Deriving the general expression for the information dimension from Eq. (3.1), one obtains

$$D_1 = \lim_{r \to 0} \frac{\sum_i -P_i(r) \log P_i(r)}{\log 1/r} \tag{3.5}$$

and the relation $H(r) \approx -D_1 \log r$ becomes apparent. D_1 describes the scaling behaviour of the partition information of the measure on the set. To clarify the idea of information here, consider that, if $P_i = 1$, i.e. it is certain to find an event in cell i, the information $-\log P_i$ gained by finding a point in that cell will be zero, while, when $P_i = 0$, the information (or surprise) will be infinite. The sum $\sum_i -P_i(r) \log P_i(r)$ in Eq. (3.5) thus measures the average information (or also information entropy) conveyed by knowing what cell a point is in. $H(r)$ assumes its maximum when the different states are all equally probable, i.e. when $p_k(r) = 1/(r^2+1)$ independent of k (for this $\sum_{k=0}^{r^2} p_k(r) = 1$ is required, thus one must remove non-unique points from the data set).

The coarse graining is carried out by varying r over an appropriate range. To be able to compare H for the different resolutions, the entropy value is normalised by its maximal value $H_{max}(r) = \ln(r^2+1)$, leading to

$$H^*(r) = H(r)/H_{max}(r) .$$

The characteristic length r^* where $H^*(r)$ has its maximum is called the entropy optimum length in paper 1. It is at this resolution were the image reveals most information and at which it should subsequently be analysed. Here the entropy analysis thus determines an optimal scale at which the support of the data shows maximum disorder while the multifractal analysis aims to describe scale independent properties of measures associated or derived from the data. In this analysis the value of H^* was found to depend strongly on the number of data points while r^* showed no such dependency. r^* was also independent of the number of randomly chosen reference cells.

The determination of r^* hence really is a robust additional means of unambiguous characterisation of point sets. In particular it avoids the asymptotic evaluations necessary in the determination of fractal dimensions, it does not involve the sometimes difficult selection of a scaling region and the analysis is feasible even for small data sets. Due to the high accuracy in the determination of r^*, even small differences in the result can be regarded to be significant. Therefore the method can also, as will be shown below, complement a multifractal analysis in that it yields information about the lacunarity, i.e. the more or less periodic occurrence of voids in the support.

3.8 Entropy Results for Hokkaido and Tohoku

To rule out possible effects due to artificial differences between the data sets such as digitisation with different resolutions, four arbitrary regions of 50 km by 50 km of the less sparse Tohoku data set were analysed instead of comparing Hokkaido and Tohoku. The locations of these regions were already indicated in Fig. 3.1. The sets will be labelled T1 to T4 in the following. The results are given in Fig. 3.11 and they are summarised in Table 3.1 together with the results of a sphere-counting analysis. The results for D_2 together with the scaling region of 0.78 km to 14.49 km explicitly confirm the fractality of the subsets. The number of events varied between 5254 and 2545 and is clearly reflected in the magnitude of H^*. r^* varied between 4.4 km and 6.4 km. Besides the differences in r^* however, the shapes of the entropy curves vary considerably which cannot be attributed to the different number of data points.

Table 3.1. Results of entropy analyses of the data sets T1 to T4 together with the correlation dimensions of the sets

Data Set	Data Size	H^*_{max}	r^*	D_2
T1	5254	0.58	5.1	1.72 ± 0.03
T2	3248	0.54	4.4	1.60 ± 0.07
T3	2545	0.50	6.4	1.72 ± 0.06
T4	2935	0.51	5.2	1.71 ± 0.05

To be able to interpret these differences, some additional calculations with artificial data sets were carried out. Figure 3.12 shows, from top to bottom, the configuration entropy curves for an original Sierpinski triangle, a

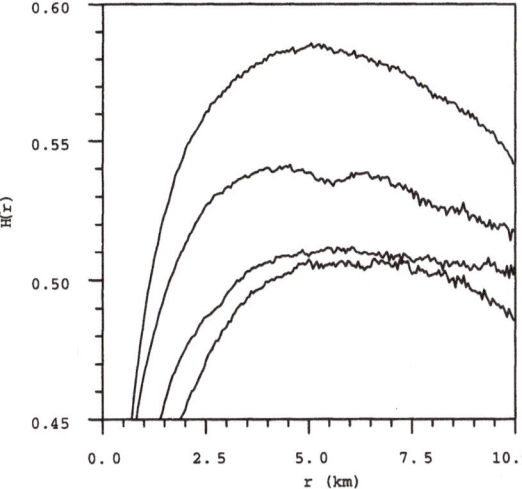

Fig. 3.11. Configuration entropy curves for the Tohoku subsets T1, T2, T4 and T3, respectively, from top to bottom

homogeneous random set (the x- and y-co-ordinates were selected at random from a gaussian distribution) and a "crystal" (a set of points placed on a an equidistant grid with $\Delta x = \Delta y = 0.69$). The sets were scaled to have an extent like the Tohoku subsets and contained the same number of data points like T1. Thus in this case also the values of H^* are significant.

The most striking result is the one for the "crystal". The low value of H^* shows, as was to be expected, a low degree of disorder and the peaks appear when r extends beyond the voids. Their spacing of about 0.69 thus corresponds exactly to the spacing of the grid points. This example can be regarded to represent an extreme case of lacunarity in that the entropy shows abrupt changes as r is varied. One may conclude that multiple peaks in H^* indicate lacunarity in general. This finding makes the entropy analysis a useful complementary tool for a multifractal analysis as lacunarity leads to piecewise linear regions in the log–log plot (e.g., [BMPV93]) which might easily be misinterpreted to be multiscaling. The curve for the random data shows no clear peak at all and the data has a relatively higher degree of disorder, both of which were to be expected. There are no multiple peaks because no roughly periodic voids are present and there is no single sharp peak because the image possesses about the same information at all scales once a minimum resolution has been reached. The Sierpinski triangle produces a somewhat similar curve to the random one in that it shows no clear maximum, rather it stays at a relatively high level once a resolution of about 4 has been reached. This is to be expected for a fractal, which by definition contains the same statistical information at all scales due to its self-similarity. However, fluctuations around this maximum level occur and again they can be attributed to the lacunarity produced by the void triangles in the Sierpinski construction. Why the overall level of H^* is higher than that of the random distribution is not clear at present but may probably be attributed to the finite sample. Care is necessary even when comparing H^* of different distributions of the same number of data points. Recalling that the Sierpinski triangle is a monofractal and that the random distribution is homogeneous as well, an important conclusion is that monofractals should not produce sharp peaks in H^*.

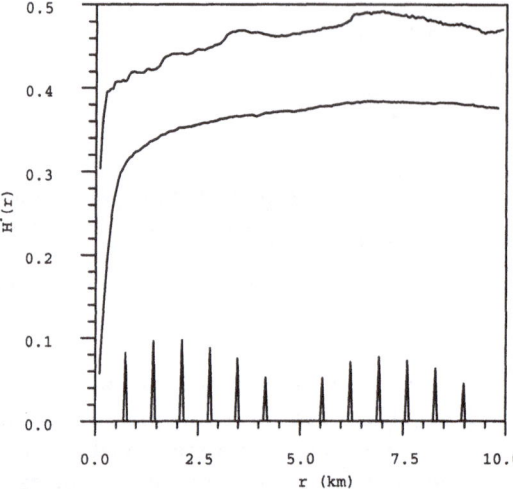

Fig. 3.12. Configuration entropy curves for, from top to bottom, an original Sierpinski triangle, a homogeneous random distribution and a regularly spaced set of grid points

Regarding the results for the Tohoku subsets again in this light and keeping in mind that the fractality of these distributions has been shown earlier, one may conclude that the comparatively sharp peaks in H^* are due to the inhomogeneity of the data. The entropy curves clearly reflect the different degrees of clustering present in each of the respective landslide distributions. The entropy analysis may thus be regarded as an indicator for multifractality of the probability measure. Comparing the curves for T1 and T4, one might also conclude that T1 is more inhomogeneous than, for example, T4 which comes close to a monofractal. The crucial advantage of this approach is that the entropy analysis can be carried out reliably for small data sets while a multifractal analysis requires much larger data sets. In the present case no sufficient scaling behaviour could be obtained over a sufficient range of q to produce multifractal spectra for the subsets T1 to T4. The double peak in the entropy curve for T2 furthermore indicates the presence of big voids with approximately periodic structure, i.e. lacunarity, which should be taken into consideration when attempting to determine the slope in a scaling analysis.

It is thus recommended to carry out an entropy analysis in addition to a (multi)fractal analysis even when the number of data points is sufficient to allow a reliable scaling analysis. Another possible benefit from the suggested procedure is that if the entropy curve shows no multiple peaks but the log–log plot is piecewise linear, the assumption of multiscaling is supported. In the latter case the entropy curve would be expected to be discontinuous at the crossover points instead of showing multiple peaks but this has yet to be numerically confirmed. It is concluded from the above findings that the entropy curves in Fig. 3.11 further support the assumption of monoscaling multifractality of the landslide probability fields as reported in Fig. 3.10.

3.9 Discussion of Landslide Results

This work has led to the unambiguous identification of the landslide process to produce multifractal measures. While the multifractal spectra for the distribution of landslide sizes and locations in Hokkaido were quite different, as was to be expected, the spectra of the probability measures for Hokkaido and Tohoku were found to be rather similar. It is not clear at present whether this should be interpreted to reflect universality of the landslide process, i.e. independence of many of the geological and meteorological details, or if it should be ascribed to the inability of the multifractal analysis to distinguish different finite data sets. A comparison with further landslide distributions of different regions from within Japan and the world would help to clarify this and is, together with a similar analysis of the earthquake epicentre distributions, intended in the near future.

While it was found that the utilised method of multifractal analysis was not suitable for local landslide distributions, regions of up to 5000 events in this case, a configuration entropy analysis was able to describe and distinguish such small sets very clearly. Beyond the possible applications within practical disaster mitigation, relations of the entropy curves with multifractal properties of the fields were also discovered. Further research in this direction seems rewarding as the problem of small data sets is frequently encountered within geoscience.

Although the recent years of research have unveiled that fractality in nature seems to be more the rule than the exception, this is not necessarily true for multifractality. Multifractality requires that measures on fractal or nonfractal supports possess similar irregularity at all scales. While multifractality in nature implies a nonlinear underlying process because no linear process is capable of producing truly intermittent signals, not all nonlinear processes produce multifractals. However, it has become increasingly clear that scaling nonlinear dynamic processes lead to multifractals on fractal supports. Multifractality does not necessarily imply chaos though, despite the fact that the invariant probability measure (natural measure) on strange attractors seems to be multifractal ([GP83], [HJK+86], [Sch88]).

The above empirical finding of multifractality thus assigns the landslide process to the same class of nonlinear processes as earthquake dynamics and rain, the main triggering mechanisms for landslides. As for disaster mitigation, the findings require a nonlinear modelling and explain the difficulty of landslide prediction which remains, as with earthquakes, impossible to date.

In particular it was shown that the prediction of the size of landslides should be considered even more difficult than the prediction of the location.

4. Some Fractal Properties of Earthquakes

"Earthquakes have so many different fractal properties that they belong to the most interesting fractal phenomena" [Tak90].

Omori discovered scaling in earthquakes in the frequency distribution of aftershocks over a hundred years ago (1895). His discovery was not accepted for a long time because for modelling purposes, a Poisson distribution was convenient([Man83]). Omori's formula:

$$n(t) \propto t^{-P} \tag{4.1}$$

says that the number of aftershocks $n(t)$, measured at time t after time t_0 of the main shock, decays following a power law. It implies a scaling property in the relation between the mainshock and its excited aftershocks ([KK78, Oga88]. Verification of Omori's formula was done by laboratory experiments on acoustic emission (AE) as well ([Sch68, Hir87a]). Aftershock behaviour is a transient fractal property in time, called the long time tail. It occurs for example in the relaxation of most amorphous materials: A simple relaxation is exponential, but amorphous materials contract slower, following the power law $t^{-\gamma}$. They possess long time correlations which means that the process depends on its history (e.g. [Tak90]).

Another well known "fractal" property is the *Gutenberg-Richter relation*:

$$\log N(m) \propto -bm. \tag{4.2}$$

where m is the magnitude and $N(m)$ is the number of earthquakes with magnitude bigger than m. Equation (4.2) may be rewritten as

$$f(m) \propto m^{-D} \tag{4.3}$$

in power-law form where $f(m)$ denotes the frequency of earthquakes above magnitude m. As this is a power law of the form (2.4), it is easy to see that D is a kind of fractal dimension. The b-value of the Gutenberg-Richter relation is theoretically related to the fractal dimension of the fracture size distribution, respectively the epicentre distribution ([Aki81, Kin83, Hir89a], see also section 7.2).

In the vicinity of Japan, earthquakes with magnitude bigger than 6 occur about 7 times a year, earthquakes with magnitude bigger than 5, about 70 times. For earthquakes with magnitude greater than 1, the relation yields roughly 530 000—on average one earthquake every minute. It is important to note that "on average" may only be taken in the strict statistical sense, as a scaling relation like (4.2) explicitly shows that there is no characteristic time span between events and that no specific magnitude is preferred. Also no characteristic fracture size exists and the size-distribution is self-similar. Thus the statistic average (recurrence period) of earthquakes does not imply a regular cyclicity (e.g. [Hsü92]): The fractal relation implies that the time interval between two events of the same magnitude is proportional to the magnitude of the event. In disaster science, the average time interval has been called the waiting time. Clustering (in space and time) has been generally found to be the rule, not the exception (see section 7 for more on this).

Using the fact that magnitude is approximately proportional to two-thirds of the logarithm of released strain energy ([Tak90] and section 7), another "fractal" relation is given by (4.2):

$$N(E) \propto E^{-2b/3} \tag{4.4}$$

where E is the total seismic energy released by the earthquake.

In the strict geometrical sense, neither D of Eq. (4.3) nor the exponent $-2b/3$ in Eq. (4.4) can be called a fractal dimension because magnitude, energy and most of the quantities mentioned here are not measured by length. From the viewpoint of dynamic systems or numerical analysis, however, these quantities may be thought of as points or vectors represented in an embedding geometrical space (just like time series are usually represented in a plane as a curve of points in a diagram of one quantity versus time).

One of the main objects of analysis in this work is the spatial distribution of earthquakes. The fractal dimension of the worldwide epi- and hypocentre pattern (i.e. the fractal support of earthquake fields) is given to be in the range of 1.2 to 1.6 by several authors in the 1980's [KK80, S+84, OA87, ASB87]. The results of these monofractal analyses were mostly estimated using the correlation function or a box-counting algorithm, the scaling region was typically between 5 and 500 km. Interestingly, D was found to be smaller (1.5 - 1.6) for deep earthquakes (280 - 700 km) than for intermediate earthquakes (70 - 280 km, 1.8 - 1.9).

Kagan (1992) found the fractal dimension of brittle shear fracture of rocks to be 2.20 ± 0.05 and, using further arguments, concludes that the conventional model of earthquake hypocentres occurring on isolated plains (i.e. $d = 2$) must be abandoned in favour of non-planar fractal fault zones. One year later, the same author goes one step further (in [Kag93]) and states that "...earthquakes do not occur on a single (possibly wrinkled or even fractal) surface, but on a fractal structure of many closely correlated faults", essentially implying that it is impossible to define individual faults. Kagan further illustrates his view by saying that "...the objective selection of fault segments is as impossible as it is infeasible to devise a computer algorithm that would subdivide a mountain ridge into mountains or a cloudy sky into individual clouds". A fractal surface in three-dimensional embedding space, however, may have a dimension ranging from 2 to 3, so the wording in the first quote is a little unfortunate.

As most earthquakes nevertheless occur along structures usually called fault zones, it is not surprising that faults, or, more generally, fractures and their distribution in heterogeneous materials, also possess a fractal structure. Indeed, the geometry of fractures in rocks and in the crust is a classic example of a fractal which holds well over a wide scaling region: It spans 10^{-6} to 10^5 m ([BS85, SA86, OA87, SB89, Hir89b]). Obviously, the scaling region and also the dimension determined depend on the resolution and detail (to what degree sub-faults are included and known) of the fault map. Results based on such maps therefore must certainly be regarded as rather crude approximations. Bodri[Bod93] showed a very good correlation between the fractal dimension of faultlines and the epicentre distribution in the Izu Peninsula, Japan. Like in the case of Omori's formula, these findings have been reproduced in rock fracture experiments ([HSI87, Hir87b].

The Omori law only states temporal scaling for aftershocks but fractality of the temporal distribution of earthquakes is a general feature (e.g. [GC95]). The spatial distribution of earthquake size has also been shown to scale (e.g. [HIY92, HLS$^+$94]).

Omori's law and the general empirical finding of temporal scaling, together with the fractality of the earthquake locations (and their corresponding spatial density), shows that the earthquake process is a scaling space-time process (cf. section aftershocks).

Sato [Sat88], by relating the values of fractal dimensions of the hypocentre distribution and the distribution of absorbers or scatterers, explains the linear relation between the logarithms of maximum amplitude and hypocentral distance of seismic waves. Thus the characteristics of seismic waves can be described as a function of hypocentral distance using the fractal dimension of the absorbers or scatterers. Also the power spectrum of seismic waves often obeys an inverse power law in a certain frequency range, indicative of scaling.

Epicentres and hypocentres are clustered, i.e. their distribution and the distribution of seismic energy is non-uniform. If a distribution has a heterogeneous fractal structure, a multifractal analysis is required to resolve it. Multifractality has also been confirmed by some authors in recent years. Their results and the implications of multifractality are mentioned in [Gol96] and discussed in more detail in chapters 6 and 7.

5. The Hurst Phenomenon

5.1 Theoretical Background

Many models and methods of analysis (using random selections) are based on two popular dogmas of science (cf. also [Kor92]):

- *Natural systems have short time memory:* The effects of random perturbations of duration τ decorrelate exponentially with $\exp(-t/\tau)$, i.e. the power spectrum will fall off for $\omega(>> 1/\tau)$ like ω^{-2} (as is the case with Brownian signals). At low frequencies, the lack of long-time correlations produces a white spectrum.
- *Small random perturbations cause predictable or neglectable changes in the future behaviour of the system.*

Particularly in geoscience, several phenomena however display unexpected long-time correlations (long-time memory, long-term persistence) and extreme sensitivity to initial conditions. The former effect is called the Hurst phenomenon, the latter is due to nonlinear dynamics, which is the subject of the second part of this work. In both cases fractals have an important role to play.

Hurst ([Hur51, Hur56]) found that fluctuations in Nile River outflows displayed unexpected long-time correlations when he wanted to determine the optimal storage capacity of the Aswan Dam (for modern references see [Man83, Man65, Kle74, Boe88]).

Here I will restate briefly the original problem to introduce some of the terms used later. Following Feder (1988), if X_t is the net inflow of water into the dam in year t, $t = 1, 2, \cdots, N$, the average inflow over N years is $\bar{X}_N = (1/N) \sum_{i=1}^{N} X_t$. If every year during the N years $\bar{X}_N$ is released from the reservoir, at year t the dam will contain

$$X(t, N) = \sum_{i=1}^{t}(\bar{X}_i - X_N) = \left(\sum_{i=1}^{t} X_i \right) - tX_N.$$

Thus, after N years, $X(N, N) = 0$.

Hurst's problem was that the dam should never overflow or go dry during N years, i.e. it's capacity had to be larger than the difference $R(N)$ between the maximum and minimum amounts of water, where

$$R(N) = \max_{1 \leq t \leq N} X(t, N) - \min_{1 \leq t \leq N} X(t, N). \tag{5.1}$$

$R(N)$ can be determined for any sequence of (random) numbers and is called the *range* of X_t. Obviously R depends on N and the standard deviation $\sigma(N)$ of the data (which grows with N).

To exclude the effect of N, Hurst ([Hur51]) introduced the dimensionless *rescaled range*:

$$R^*(N) = \frac{R(N)}{\sigma(N)}. \tag{5.2}$$

A simple experiment in which k coins are tossed N times and where X_t is defined to be the difference between the number of heads and the number of tails at experiment t yields: $R^*(N) \propto \left(\frac{\pi}{2}N\right)^{1/2} (\sigma = k^{1/2})$.

But when Hurst *et al.* ([Hur56]) analysed 120 geophysical and statistical time series, they found

$$R^*(N) \propto N^H \tag{5.3}$$

with $H = 0.73 \pm 0.09$, i.e. significantly different from the theoretical 0.5 (The relationship $R(N)/\sigma(N) \propto N^H$ is meant when *Hurst's rule* is mentioned. Thus also the name R/S analysis for a Hurst analysis).

Some possible reasons for such a behaviour could be ([Boe88]) non-stationarity, pre-asymptotic behaviour or long-time correlations in the stationary X_t.

If the rescaled range is calculated analytically for independent standard normal variables ([AL76]) and the *local Hurst exponent*

$$H_N = (\log R^*_{N+1} - \log R^*_N)/(\log(N+1) - logN)$$

is plotted as a function of N, H can be seen to asymptotically reach 0.5 at about $N = 10000$. For correlated sequences with limited time dependency, H approaches it's theoretical value much later (see [Fed88] for a computer simulation of Hurst's original "biased card" experiment). Thus only stationary processes with auto-correlations falling off slowly as a power of the time lag can preserve a $H > 0.5$ for $N \to \infty$. Fract(ion)al Brownian motion (fBm, see e.g. section 9.3.3) represents such data.

A continuous process $y(t)$ is a Brownian process (a continuous-time random walk) if, for any time step Δt, the differences $\Delta y(t) = y(t + \Delta t) - y(t)$ have the following properties:

1. They are Gaussian (i.e. $y(t)$ is Gaussian).
2. Their mean is 0.
3. Their variance is proportional to Δt, which means, due to item 2, that successive differences are uncorrelated.

The generalisation to fractal Brownian processes is done by simply replacing property 3 by

– Their variance is proportional to Δt^{2H}

where $0 < H < 1$ is the Hurst exponent (so called by Mandelbrot in [Man77]).
Ordinary Brownian motion has $H = 0.5$, i.e. it is uncorrelated white noise.
Again property 3 is equivalent to a correlation property of successive increments: In the fractal case, successive increments are correlated with a
correlation coefficient ρ which is defined:

$$\rho = \frac{2^{2H} - 2}{2}, \quad -1/2 < \rho < 1. \tag{5.4}$$

This correlation is independent of the time step and, together with properties 1 and 2, characterises the scaling behaviour (thus it's fractal dimension)
of a fractal process: If $y(t)$ is a fractal process with Hurst exponent H, then,
for any constant $c > 0$, the process $y_c = (1/c^H)y(ct)$ possesses the same
statistics (thus, one might require for $y(t)$ to be a fBm: $y(t)$ is statistically invariant with respect to the affine transformation $t \to ct, y(t) \to c^H y(t)$ which
shows that these processes are self-affine rather than self-similar). This property is exploited in the renormalisation approach (the rescaling by c is called
renormalisation) (cf. also [Tur92]). Another important practical application
resulting from the above is the fact that

$$D = d - H \tag{5.5}$$

where d is the Euclidean embedding dimension of a graph ((hyper-)surface),
D is its fractal (capacity) dimension and H was determined from the trace
of the graph in $d - 1$-dimensional space (see e.g. [HS93]). One may thus determine D of any graph in three dimensions (e.g. a geophysical field) by estimating H for profiles in arbitrary directions. The latter possibility becomes
especially interesting for anisotropic (possibly self-affine) data and will be
exploited further below.

The correlation between y_1 and y_k for a fBm is (from the scale invariance
and Eq. (5.4) thus proportional to $\sigma^2 H(2H - 1)k^{2H-2}$ if $k \gg 1$ and $H \neq 1/2$.
The power spectrum of a fBm with Hurst exponent H is hence $G(f) \propto$
f^{-2H+1} (self-similarity is achieved for $H = 1$ and the power spectrum reduces
to $G(f) \propto f^{-1}$). H is thus simply related to the exponent of the power-law
autocorrelation function of $y(t)$.

Furthermore, a power law fall-off of the power spectrum of a fractal curve
(e.g. a fBm) is expected ([Vos88]): the power spectrum should fall off $\propto f^{-\beta}$
with $\beta \approx 2H - 1$. This property makes possible an independent check for the
determination of H (respectively a determination of H by getting β or vice
versa, see e.g. [HS93]). Also the power spectrum of a fractal must obviously
be broad because no characteristic scales exist.

Random processes with $1/f$-like power spectrum are called $1/f$-noise or
flicker-noise or also pink noise. Contrary to the extremes of white noise

$(G(f) \propto f^0)$ and Brownian noise $(G(f) \propto f^{-2}$, the integral of white noise, cf. also section 9.3.3) $1/f$-noise was observed to be the "most natural" for modelling natural (geophysical) processes (see for example [CJ89b] for an application in refraction seismics). $1/f$ spectra were concretely observed from time series such as for example the velocity of undersea ocean currents, solar normal modes, sunspot numbers and weekly earthquake frequencies. The above mentioned analysis of the secular motion of the earth's pole ([MM70]) also showed this behaviour. In the same paper, a further important property of $1/f$ processes is mentioned: Due to the large amount of energy contained at low frequencies ("red" spectra) the processes are invariant under averaging, i.e. averaging (smoothing) cannot reduce the randomness of the signal. A possible explanation for $1/f$ behaviour comes from the theory of self organised criticality.

Flicker noise and the Hurst phenomenon pose the same puzzling question: How can a system (an electric resistor in the case of electronics, the temperature at a specific geographical location in the case of meteorology) carry on the influence of a "local" trend over very long time intervals? "By what sort of physical mechanism can the influence of, say, the mean temperature of this year at a particular geographic location be transmitted over decades and centuries?" ([Kle74]).

Problems related to the interpolation or prediction of processes exhibiting the Hurst phenomenon (see for example [MN68, Hew86]) yielded the following interpretations of the value of H:

- For $H = 1/2$, successive steps are independent and the best prediction is the last measured value.
- For $H > 1/2$, the local trend over the interval will continue (frequently termed *persistence*). The best prediction is based on an extrapolation of the trend.
- For $H < 1/2$, the local trend will reverse (*antipersistence*). *The best prediction tends to the mean value over the interval.*

Mandelbrot and Wallis (1968) introduced the terms "Joseph effect" to designate persistent processes with $H > 1/2$ (after *Genesis*, 41.29–30, where it is observed that seven plentiful years shall be followed by seven years of famine) and "Noah effect" for processes with $H < 1/2$ (after *Genesis*, 7.11–12, where very rapid and large (intermittent, singular) events happen).

Recent applications of R/S analysis which yielded $H > 0.5$ include the secular component of the earth's polar motion ([MM70]), precipitation data ([MW69, HS93]), global climatic change (oxygen isotope ratios) ([FS89]), reservoir performance (density porosity logs) ([Hew86]).

5.2 The Hurst Phenomenon and Earthquakes

Besides the fact that weekly earthquake rates show persistence, in that the time series possess a Hurst exponent $H > 0.5$ (see above), an early example of "Hurst analysis" is shown in Fig. 5.1: In this plot from Richter (1958) , the cumulative global strain release from large shallow earthquakes from 1904 to 1955 is shown. The plot strongly resembles the Devil's staircase, a plot of the cumulative mass of the Cantor bar (see e.g. [Fed88]), showing that an underlying distribution of earthquakes of a particular H exists. The latter in turn already shows that, even without knowing the value of H, the earthquake process possesses long time memory. It is interesting to note that Devil's staircases frequently arise in physical systems with non-linearly coupled oscillatory components (see [Bak86]), leading also to "clusters of activity" which can easily be observed in Fig. 5.1.

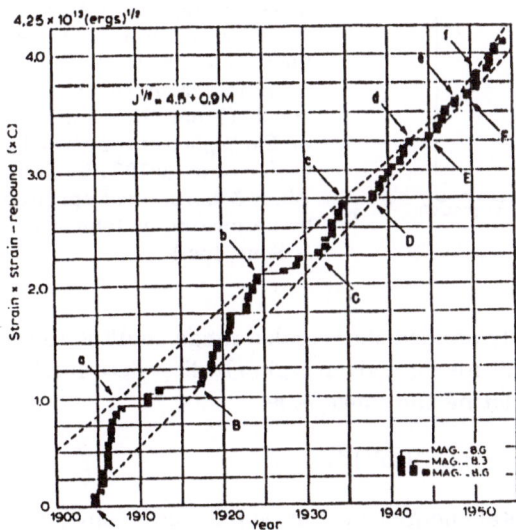

Fig. 5.1. Strain release from shallow large earthquakes on a worldwide basis exhibiting the typical shape of a Devil's staircase (From *Elementary Seismology* by Charles F. Richter. ©1958 by W. H. Freeman and Company. Used with permission.)

A thorough treatment of the Hurst phenomenon in connection with earthquake cycles (i.e. clustered activity, not periodic behaviour) may be found in Lomnitz (1994) where the Hurst method is used as a model for the prediction of the nonlinear clustering process of earthquakes (quiet years tend to be followed by quiet years and active years by active years). Figure 5.2 gives a Hurst diagram of cumulative earthquake energy for Mexico, where the large dot represents the 1985 Mexico earthquake. The upper and lower dashed lines indicate the range $R(N)$ as determined according to Eq. 5.1, D signifies the deviation of the cumulative energy function from its average trend. One immediately recognises the resemblance to Fig. 5.1 in that both curves constitute a Devil's staircase.

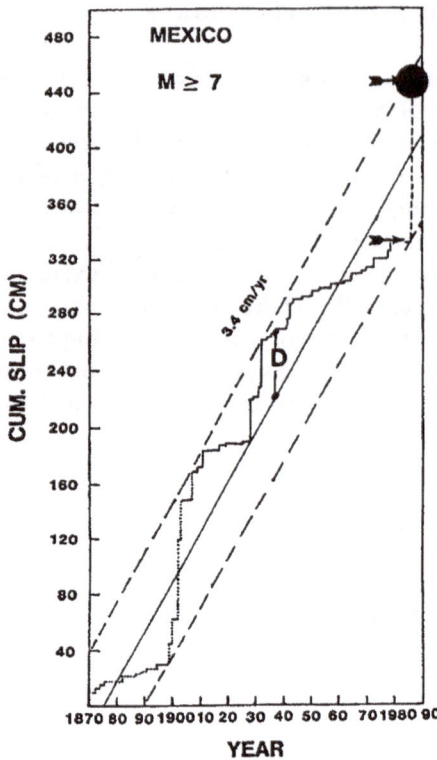

Fig. 5.2. Cumulative earthquake energy for Mexico in a Hurst diagram. Dashed lines indicate R, D is the deviation from the average trend (From *Fundamentals of Earthquake Prediction* by C. Lomnitz. ©1994 by John Wiley & Sons, Inc. Reprinted by permission of John Wiley & Sons, Inc.)

Lomnitz (1994) has successfully applied the rescaled range analysis to estimate the future maximum magnitude for a given region by first determining H for that region and then calculating the range R as

$$R(N) = \sigma(N)(aN)^{H}.$$

R here gives an estimate of the total energy that can be released in any given time period N. If σ is assumed to be constant, a conservative estimate of the upper bound M of the maximum magnitude can be obtained from

$$M(N) = 0.75 \log R(N) - 3.$$

Ogata and Abe (1991) applied a Hurst analysis to the magnitudes of earthquakes in Japan and the world and obtained values of H of about 0.5, apparently signifying independence of earthquake events. As pointed out by Lomnitz (1994) however, the analysis was meaningless due to two reasons: the unphysical magnitude should not have been used as a measure and the Hurst analysis should not be applied to large complex regions, where localised effects superimpose in such a way that the statistics is destroyed (see also the discussion of homogeneous seismogenic structures in appendix B).

5.3 Application of Hurst Analysis to Seismicity

While the idea of the Hurst phenomenon and analysis originates from time series analysis, the principles can easily be applied to any "profiles", e.g. for determining D (see Eq. (5.5)). In practice, different approaches exist for the estimation of H, a good overview including program listings is given by Hastings and Sugihara (1993). The approach based directly on Eqs. (5.1) and (5.3) respectively (5.2) and (5.3) is illustrated in Figs. 5.3 to 5.5: Fig. 5.3 shows a portion of a profile with the lower and upper bounds (R in Eq. (5.1)) for a neighbour distance of $\pm$ 2 points. The greatest vertical difference between the bounds becomes one point (R) in a $\log N$ versus $\log R(N)$ plot from which H can be estimated by linear regression according to Eq. (5.3). The procedure is shown in more detail in Fig. 5.4, where a fractal elevation

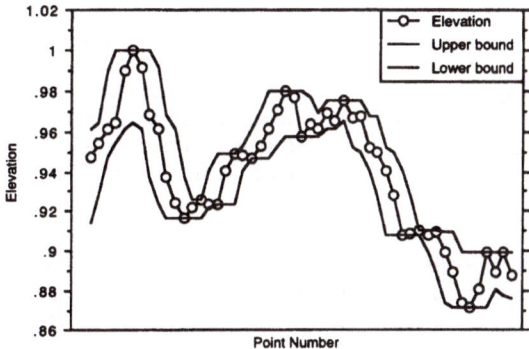

Fig. 5.3. Upper and lower bounds of the range of a fractal profile at 2nd neighbour distance (from Russ, 1994)

profile and a zoom window containing the interval with the largest 1st neighbour and 4th neighbour ranges is shown. Thus 2 points for the Hurst plot are obtained. The resulting complete plot is given in Fig. 5.5 which was obtained by varying N over an appropriate range of Nth nearest neighbour distances. The slope of the log–log plot gives H and therefore, using Eq. (5.5), D.

Also from Fig. 5.3 or 5.4 it becomes clear that a single outlier can ruin the Hurst statistics—the method is very sensitive to noise. Obviously the greater the dynamical range of the signal, the greater the associated range of error. In the case of seismic energy for example, the application of the Hurst method outlined above is questionable: When converting the magnitude values given in earthquake catalogues into seismic energy or seismic moment by the formulas given in section 7.2, an error of ± 0.1 in the magnitude value translates into an error of e.g. -29% to +41% in the energy (cf. [Lom94]). Below, I therefore use the rupture size as a measure for earthquake size (see section 7.2).

Here, however, the interest lies in the determination of H as a function of direction as a means to detect and characterise anisotropy of the fractal

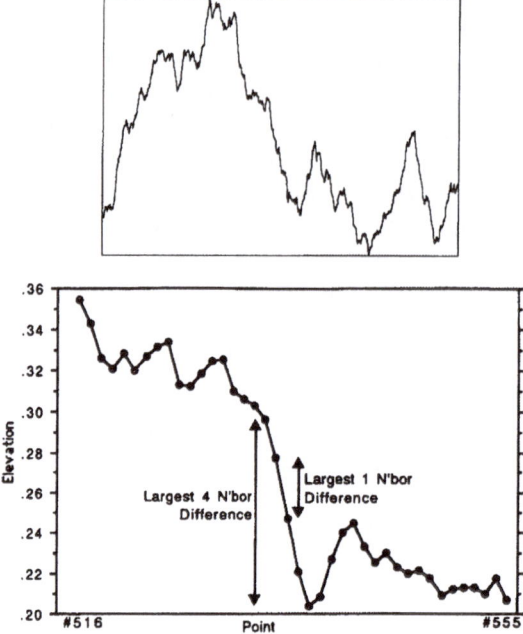

Fig. 5.4. A fractal elevation profile and an enlarged window showing the greatest range for 1st and 4th nearest neighbours in the whole record (from Russ, 1994)

dimension. An obvious and easy approach in the case of epicentre distributions would be to simply calculate the fractal dimensions of "Cantor dusts" describing the distribution in one direction, i.e. of sets of points obtained from the intersections of the epicentre distribution with a line. Such analyses have been carried out by e.g. Hubert and Carbonnel (1991) in the case of tropical rainfall data obtained over an area of over $10\,000$ km^2 by employing 111 rain gauges. The authors conclude that the rain fields possess anisotropic fractal properties and suggest that the anisotropy might be controlled by the direction of movement of meteorological perturbations (see e.g. [SL93] for a detailed description of anisotropy in the case of atmospheric phenomena). The disadvantage, especially in the case of sparse data like epicentre distributions, is the poor statistics obtained by this method because it only uses points along single profiles.

A statistically more sound approach suitable for two-dimensional arrays of a measure (i.e. geophysical fields or, e.g. in materials science, a matrix of surface elevation data) is to move a disk of radius r across the data and to record the differences between highest and lowest values within the disk (see [Rus94]). This, however, is only applicable to isotropic fields because the range is determined regardless of direction (the Hurst plot would only report the highest fractal dimension encountered in all directions). A generalisation which allows to record the range as a function of distance and direction is called the Hurst Orientation Transform (HOT). By comparing the values of

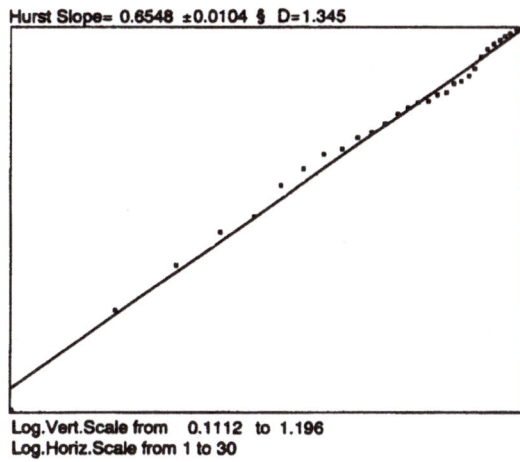

Hurst Slope= 0.6548 ±0.0104 § D=1.345

Log.Vert.Scale from   0.1112 to 1.196
Log.Horiz.Scale from 1 to 30

Fig. 5.5. Hurst plot ($\log N$ vs. $\log R(N)$) for the profile in Fig. 5.4, giving H as the slope (from Russ, 1994)

the measure for all pairs of values within the Hurst disk of radius r which is in turn moved across the field systematically, a table containing the greatest differences as a function of orientation and distance is constructed. From the HOT table, H can be obtained as a function of orientation to reveal the possible anisotropy ([Rus90]). Here, a slightly modified approach had to be used due to the sparsity of the earthquake data: Differences between occupied and void cells of the fields constructed as input for the Hurst analysis (cf. chapter 7, fields were generated in the same way as for multifractal analysis) had to be neglected because otherwise the great steps between void and occupied would overwhelm the relevant statistics of the areas where seismicity occurred. Sparsity of the earthquake fields also prevents the use of Fourier analysis for an anisotropy analysis because there is no way to interpolate the gaps[1].

The HOT table itself can be plotted, using brightness as a measure for the ranges (see below). From the HOT table, a Hurst plot is constructed from each row corresponding to one particular direction. The number of points in the table and thus also the number of distinct directions depends on the size of the Hurst disk (the larger the disk, the finer the spatial and angular resolution). In several experiments, a size of $r = 16$ was found to be sufficient for the fields of 256×256 cells used throughout this work. Obviously, the number of points on the individual Hurst plots also varies with direction, so that unfortunately some values of H are more reliable than others. However, as the results are only used to compare data sets and no absolute accuracy is aimed at, this drawback can be safely ignored.

---

[1] To a certain extent, it might be possible to use Iterated Function Systems to fractally interpolate the measure (e.g. [Bar88]) before performing a Fourier analysis.

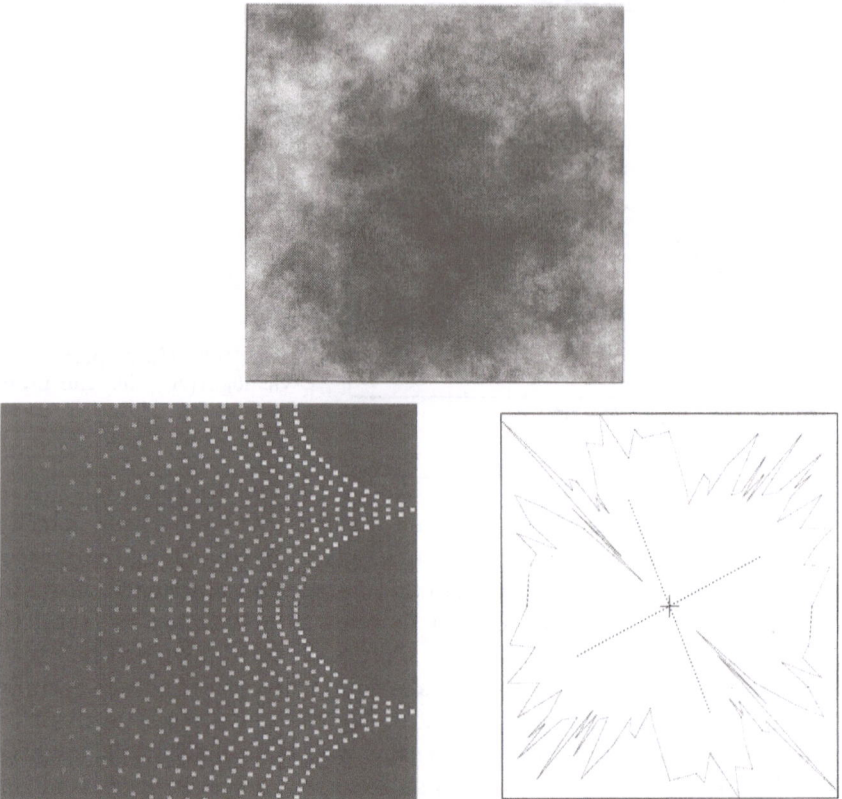

Fig. 5.6. Range image of an isotropic field ($D = 2.5$) with resulting HOT table and rose plot of H as a function of direction (see text)

One example each of an isotropic and an anisotropic field are given in Figs. 5.6 and 5.7. Shown are the fields as range images, i.e. as a gray-scale rendering where brighter pixels represent higher values, their respective HOT matrices (for clarity each position has been enlarged to an area of $3{\times}3$ pixels) and finally their rose plots. The HOT plots show the maximal differences R as a function of neighbour distance N (horizontal axis) and direction (vertical axis) while the rose plots show H as a function of angle together with the major and minor axes of a least squares fitted ellipse from which the degree of anisotropy is estimated. Note that the degree of anisotropy is not estimated as a ratio of the extremal values of H because of the possible unreliability of the individual value of H mentioned above. The isotropic example was generated by using a fBm with $H = 0.5$ in two dimensions (see [PS88]), hence a fractal dimension of 2.5 in all directions is expected. For the anisotropic example, an FFT method was employed by first generating data with an appropriate slope in the log(power) versus log(frequency) plot in Fourier space and then retransforming it into the time domain with phase randomisation (see also [PS88] for details). The latter method would produce an isotropic fractal if not two slope values were used which produce a maximum and minimum fractal dimension orthogonal to each other. Here slope values of 0.8 and 0.1 were used, resulting in extremal fractal dimensions of 2.2 and 2.9.

The resulting D-values as obtained for the isotropic case varied between 2.30 ± 0.15 and 2.65 ± 0.07 (average 2.55), the ratio of the ellipse axes (otherwise called degree of anisotropy here) was 0.96, i.e. close to unity as expected for the isotropic case. The orientation of the ellipse, i.e. the direction of extremal fractal dimension with respect to a horizontal line (i.e. geographic North), was found to be about 25° which was in relative good agreement with the expected 30°. In the anisotropic case, a degree of anisotropy of 0.52 and fractal dimensions in the range of 1.93 to 2.92 resulted.

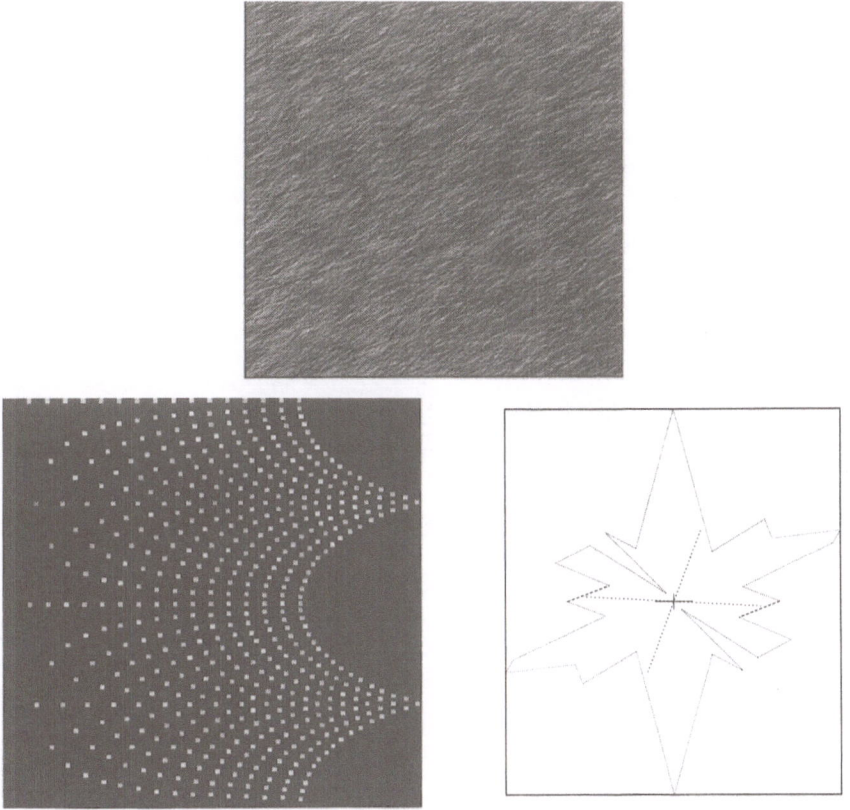

Fig. 5.7. Range image of an anisotropic field ($D = 2.2 - 2.9$) with resulting HOT table and rose plot of H as a function of direction (see text)

To conclude this section, it should be pointed out that fractal anisotropy may of course vary with thresholding the data, i.e. anisotropic multifractals exist (e.g. [SL93]). Here however, the analysis of directionality of the fractal dimension is carried out separately from multifractal analyses.

6. Multifractal Spectra as Precursors

6.1 Multifractal Earthquake Precursors

As has been pointed out earlier, (cf. also [Mei94]) a single fractal dimension is generally not very useful to characterise changes in temporal or spatial seismicity patterns and therefore quite useless in the search for possible precursors. Again, this is not surprising as the earthquake process belongs to the class of phenomena which produce multifractal measures. Seismicity patterns are therefore inhomogeneous and changes are expected to show up the clearest in changes in the degree of heterogeneity, not in properties of the support , i.e. D_0, or also D_2.

While the multifractality of earthquake patterns has been firmly established by now, only very little on the use of multifractal spectra respectively their change with time as a precursor may be found in the literature. This is so although several authors mention the exciting possibility (e.g. [HI91, GC95, HIY92]) and announce further research in this direction.

In fact only two sources are known to me where the problem is approached: Haikun (1993) presented a talk entitled "The Multifractal Local Scaling Feature of Spatial 'Energy Generating' and its Seismic Precursory Information" at the International Symposium on Fractals and Dynamic Systems in Geoscience in Frankfurt in which he discusses precursory changes of the $f(\alpha) - \alpha$ curve. Besides the theoretical study of the spatial partitioning of seismic energy, he analysed earthquake catalogues for the years 1970–1991 for the Datong area, China, for multifractal features. For all three moderately strong earthquakes in 1976, 1989 and 1991, repeatable pre- and post-seismic features in the $f(\alpha) - \alpha$ curves are mentioned in [Hai93]: *several years* before an earthquake, the curve changes from "about symmetrical or a little low with the right side to much low with the left side". Also the range in α, i.e. the opening of the upside-down parabola, became wider and the maximum, i.e. $f(\alpha_0)$, moved to the right. Haikun (1993) found by computer modelling that this behaviour corresponds to a transition from stochastic to ordered and clustered patterns and that the seismic energy distribution changed from homogeneous to inhomogeneous. He also found that this process ended 1–2 years before the main shock. After the main shock, he observed the reversed behaviour. In only one case, before the 1989 Datong-Yanggao earthquake, the multifractal spectrum showed a transition from "ordered (i.e. clustered) to disordered" during a short period.

He concludes that the temporal variations in $\Delta\alpha$ and Δf "provide some message for medium forecast". The variation in $\Delta\alpha$ is caused by a simultaneous translation of α_{min} and α_{max} while the variation in Δf is mainly caused by a changing $f(\alpha_{min})$. He explains that the latter means that the "$\Delta\alpha$ seismic precursory information" comes from both the vicinities of very small measure as well as the very high areas (i.e., for the occurrence probability measure: sparsely populated and highly clustered areas within the area of observation) while the Δf information is caused almost exclusively by large values within the measure (e.g. larger earthquake events respectively dense clusters). Finally Haikun (1993) summarises that the pre-seismically expanding $\Delta\alpha$ indicates an increase in the range of local scaling properties which implies increasing complexity and that the shift of $f(\alpha_0)$ to higher values of α shows that areas of high values of the measure increase while almost void areas expand. The latter in turn implies a rising energy concentration. Although he also mentions a change in $\Delta f = f(\alpha_{max}) - f(\alpha_{min})$, it is not mentioned which extremal value of α is the cause and whether the respective values of f increase or decrease.

Details of Haikun's (1993) analysis are unfortunately not available. It would have been important to know the spatial extent respectively the choice of seismogenic structures included, the size of the temporal window and the data sets at large as well as the absolute values obtained. The latter would have been interesting for a comparison. The problem of area choice is addressed further below. Here the question is whether Haikun analysed areas peculiar to each of the three mentioned earthquakes or whether he regarded the Datong area as a whole (the word "local" in the title of his presentation refers to the local scaling exponent α, not to local seismicity). However, in spite of these shortcomings, a schematic illustration of the above mentioned findings is given in Fig. 6.1. It allows a discussion of Haikun's (1993) interpretations, also in the light of Chapter 3. Looking at Fig. 6.1, one may summarise and add to Haikun's (1993) interpretation:

- The increase in $\Delta\alpha$ indeed means a transition from homogeneous (random, space filling) to heterogeneous (ordered, complex, clustered) patterns.
- The shift to the right of α_{min} means that the clustering within the most clustered areas becomes more intense (the local fractal dimension increases within these vicinities).
- The analogous shift of α_{max} indicates that the clustering within the sparse areas also increases. Thus the degree of clustering increases overall.
- The increase of α_0 shows that most clusters now possess a higher local fractal dimension than before. Should α_0, for the case of epicentres with an embedding dimension of 2, increase from below 2 to above 2, it would mean that before a major earthquake most clusters are surrounded by denser clusters while before most events were surrounded by sparser clusters.

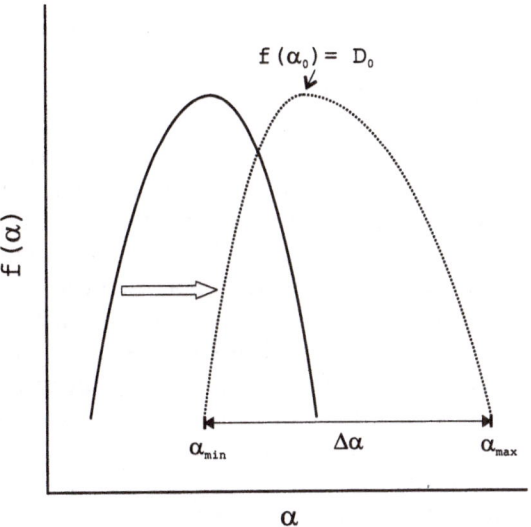

Fig. 6.1. Schematic Illustration of the precursory behaviour of multifractal spectra as observed by Haikun (1993): The $f(\alpha) - \alpha$ curve changes from the solid one to the dashed one several years before a major earthquake occurs

- As $f(\alpha)$ tells how many clusters of a given α exist, it would have been important to know whether e.g. $f(\alpha_0)$ also changed. An increase in $f(\alpha_0)$ would mean an increase in D_0—which, see above, is not usually observed.
- The mentioned increase in Δf, without further details, only tells that a change occurred in the ratio of highly clustered and sparsely populated areas.

In chapter 7, the above findings will be tested on data sets from Japan.

The second work which explicitly addresses multifractal precursors is the one by Hirabayashi *et al.* (1992). While the main focus of that paper is on showing that seismicity from three major regions (California, Greece and Eastern Japan) constitutes multifractal measures, the last section is entitled "Temporal Change of the Multifractal Distribution". There, the authors analyse the temporal behaviour of the spectrum of generalised dimensions (the D_q curve) obtained from the epicentre distributions.

The authors mention the contradicting requirements of as short as possible data sets to be able to detect sudden changes and of as long as possible temporal windows to achieve sufficient numerical accuracy. They compromise in using windows of 500 events, pointing out that the absolute values of D_q are not reliable then but that the relative changes may still be considered significant. The analysis was carried out with windows which overlapped 250 events. Important is the fact of keeping the number of events constant to ensure comparable results—it seems physically desirable to keep the time interval constant to include the temporal dynamics of the seismicity pattern but then the number of events would vary extremely due to the strong spatio-temporal coupling as shown by, e.g., the Omori law. It seems that the temporal behaviour itself has to be addressed separately as has been done in section 7.3.1.

Figures 6.2 to 6.5 show results ([HIY92]) given as examples for California and Japan: the upper part of Fig. 6.2 shows the earthquake frequency per 10 days of Californian earthquakes for 1971–1985, the lower part selectively shows the temporal behaviour for D_{-2} and D_2, i.e. the correlation dimension. Most notable is the increased activity around event number 3000 where the strongest earthquake of magnitude $m = 6.1$ occurred and the corresponding increase in D_{-2}. Again it is confirmed that D_2 alone is not useful for the distinction of anomalous seismicity patterns. Although it might seem here that e.g. D_{-2} would be a sufficient measure, Fig. 6.3 is even more convincing and yields much additional information. The left part shows a D_q spectrum for the Californian earthquakes for a "normal", quiet period (here events number 0 to 500) while the right part shows the spectrum for the period 3000 to 5000. The transition from almost monofractal to a definite multifractal is very obvious. Recalling the relationship between the D_q and $f(\alpha) - \alpha$ curve (section 2.2.6) the change shown in Fig. 6.3 may be described to be very similar to the observations by Haikun (1993) in "$f(\alpha) - \alpha$ terminology". Here, however, a significant increase in D_0 may also be observed, i.e. $f(\alpha_0)$ goes up, the density of the support increases. Unfortunately the steep D_q curve was obtained during the earthquake, not before, so its precursory value is not evident here. The change in D_{-2} shown in Fig. 6.2, however, might possess precursory quality.

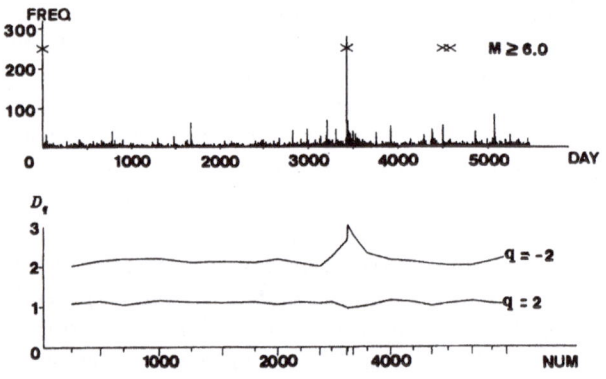

Fig. 6.2. 10-daily frequency of earthquakes in California with events of $m > 6.0$ indicated by crosses (upper part) and the temporal behaviour of D_2 and D_{-2} of the epicentre distribution (from Hirabayashi *et al.* (1992))

Looking at the lower part of Fig. 6.2, the question of how to represent, respectively compare, multifractal spectra over time comes to mind: it would certainly be as impractical to plot many $f(\alpha) - \alpha$ curves as it would be to plot the curves for all values of q as is done in Fig. 6.2 for $q = 2, -2$. Instead one might want to summarise the multifractal properties into the *non-uniformity factor*

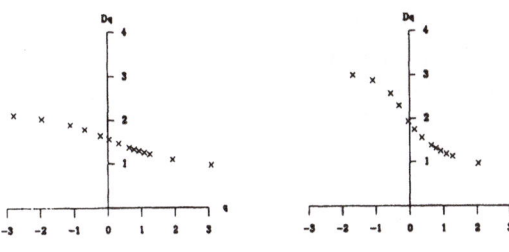

Fig. 6.3. Gentle, almost monofractal, D_q curve (left) and definitely multifractal spectrum (right) for seismically quiet and active intervals respectively of the data in Fig. 6.2 (from Hirabayashi et al. (1992))

$$\Delta = \frac{\alpha_{max} - \alpha_{min}}{f(\alpha_0)},$$

i.e.

$$\Delta = \frac{D_{-\infty} - D_{\infty}}{D_0}.$$

For a monofractal, Δ should thus be close to 0, while for a multifractal, the non-uniformity factor may theoretically assume any real value > 0. Changes of the capacity dimension of the support are taken care of by the normalisation by the denominator so that Δ is really comparable throughout different time windows. Although Δ can of course not completely describe the spectra, it will be convenient in the representation of temporal changes of multifractal properties in chapter 7 where it is also used for the first time in multifractal applications. Experience throughout this work has shown, however, that, due to largely varying scaling limits as a function of q, the lower limit for q which can still be safely determined, varies from dataset to dataset. Therefore a simplified non-uniformity factor $\Delta' = (D_{-2} - D_2)/D_0$ is determined where indicated.

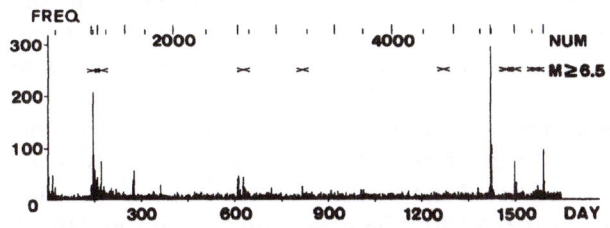

Fig. 6.4. 3-daily frequency of earthquakes in Eastern Japan with events of $m > 6.5$ indicated by crosses (from Hirabayashi et al. (1992))

Figures 6.4 and 6.5 show similar results for Japanese earthquakes of the time 1983–1987. The same phenomenon, namely the transition from gently sloped to steep D_q curves is observed although even the gentle type has a more definite multifractal character than the gentle Californian spectrum, indicating that seismicity in Japan is generally more inhomogeneous than in California. The steep spectrum in Fig. 6.5 was obtained during the interval centred on event number 556 where an $m = 7.7$ earthquake occurred. Hirabayashi et al. (1992) point out that in the case of Japan, steep D_q spectra were also found in quiet periods, e.g. period 3500-4000, indicating a worse correspondence of seismic activity and D_q curve for Japan than for California. Inspecting Fig. 6.4, however, one notices that the offending interval starts shortly after a major earthquake, thus making it possible that the reverse transition had not yet taken place (completely).

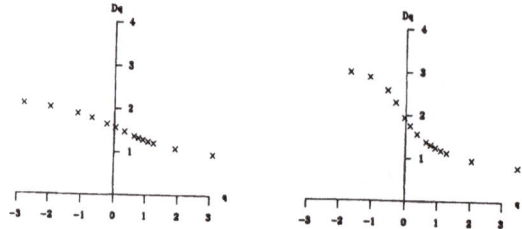

Fig. 6.5. Gentle multifractal D_q curve (left) and extreme multifractal spectrum (right) for seismically quiet and active intervals respectively of the data in Fig. 6.4 (from Hirabayashi *et al.* (1992))

In conclusion, Hirabayashi *et al.* (1992) regard the continuous observation of D_q as a promising method to detect anomalous changes of seismicity, emphasising the great sensitivity of D_q for $q < 0$. For prediction purposes, the size of the area should be restricted and the lower magnitude cutoff decreased to include microseismicity.The latter two statements are important and are addressed again further below.

6.2 Multifractal Phase Transitions

After the (semi-)empirical evidence for multifractal precursors or at least multifractal changes in connection with anomalous seismicity given in the previous section, it is interesting to take at least a superficial look at the possible theoretical reasons for (abrupt) changes in the $f(\alpha) - \alpha$ curve *before* major earthquakes:

The connection between earthquakes and turbulence was already mentioned in Hirabayashi *et al.* (1992). The first fractal model of turbulence was monofractal ([FSN78]) and could not explain the intermittent (nonlinear) fluctuations of velocities ([BPPV84]). Meneveau and Sreenivasan (1987) experimentally found the dissipation field to be multifractal by analysing one-dimensional sections of turbulent flow. During that work, it was found that D_0 was 1, e.g. there are no areas within a turbulent flow where absolutely no dissipation occurs. The capacity dimension of epicentre distributions is usually close to 2, i.e. there are no earthquake-free areas either. Revision of the physical model for turbulence led to multifractals as described in [Gol96] where the basic idea of cascade processes was introduced. Earthquakes might be regarded to be analogous to the vortices in turbulent flow where energy is dissipated.

It thus becomes obvious that earthquakes may be regarded as turbulence in solids (cf. [HIY92]). However, earthquakes are coupled in space and time, i.e. they are an integrated spatio-temporal phenomenon which is even more complicated than turbulence. Consequently a theoretical model for earthquakes must be considered elusive in the near future. Current studies focus on the implications of self-organised criticality and the properties of cellular automata.

While the analogy between earthquakes and turbulence is established by the above facts, it remains a question why the multifractal spectrum should reflect anomalous seismicity or even show precursory behaviour before a major earthquake. This question is closely related to phase transitions and therefore critical states, universality and possibly chaos. Here however, only the connection between phase transitions and the multifractal spectrum as a descriptor of the "energy dissipation field" will be given. The energy dissipation field corresponds to the distribution of seismic energy in the crust, or, as another measure of seismic activity, the spatial density of epicentres. As already mentioned before, a further analogy exists between multifractals and thermodynamics ([Ott93]) in that f and α correspond to the entropy (cf. [Gol96]) and internal energy per unit volume respectively. Certain changes in the multifractal spectrum are also referred to as multifractal phase transitions (e.g. [Hoo93] in connection with earthquakes).

A classical example of a (second order or continuous) phase transition, which will also explain the occurrence of scaling, is the transition of water to vapour. At a pressure of 1 at, the boiling point of water is 100°C. At the transition from liquid to vaporous state, the volume of a given mass expands suddenly to about 1600 times its original volume, i.e. the phase transition is accompanied by an abrupt jump in the density (the phase transition w.r.t. the density is of first order). If the external pressure is increased, the boiling point raises while the difference in density tends to 0 at the *critical point* (218 at at 647 K). At this point, the dry vapour has the same density as the boiling liquid, i.e. the two phases cannot be distinguished. Steam bubbles and water droplets are mixed together at all scales, making the substance scale invariant, i.e. fractal. *Near* the critical point, the substance becomes milky (opalescent) because the droplets reach a dimension of the order of the wavelength of light. The latter might be regarded as a macroscopic precursor which is easily observable and which is due to the divergence of correlation length at microscopic scales (which might not be observable). In the latter case, patterns change from non-fractal to fractal. Another example of "pattern change" is given by the transition from magnetic to non-magnetic in ferromagnets: as the temperature approaches the Curie temperature, the magnetisation M goes to zero (with no outer magnetic field) and the formerly parallel, i.e. ordered, elementary magnets cannot interact anymore (the correlation length diverges) and become disordered. The magnetic transition is thus accompanied by a change from order to disorder. Again a power law applies near the critical temperature T_{cr}:

$$M \propto |T - T_{cr}|^{\beta}.$$

If one wants to carry the idea of "seismic phase transitions" further, one might possibly say that the transition from one seismic pattern to another (be it from random to complex or vice versa, but significant in the singularity spectrum and hence a "multifractal phase transition") during a major earthquake is accompanied by the sudden release of an enormous amount of seismic energy (an analogous formulation can be made when regarding the energy release as the primary characteristic, then being accompanied by a sudden change in epicentre distribution). Indeed the fact that seismicity patterns seem to possess scaling at all stages of the seismic cycle supports the assumption that the earth's crust is in a constant critical state (as suggested by the theory of self-organised criticality).

To end this section, it should be summarised that phase transitions in general are singularities of some characteristic quantity (e.g. the energy) at certain critical parameter values. As the multifractal spectrum is the spectrum of singularities (earthquakes: in the epicentre probability distribution or in the distribution of seismic energy), it shows the singular behaviour near critical points. The theoretical D_q or $f(\alpha) - \alpha$ curve becomes non-differentiable (see e.g. [OGY89]). Examples of such curves for certain models (maps) showing chaotic behaviour are shown in Figs. 6.6 and 6.7. Due to numerical limitations, such shapes will never be discernible in spectra obtained from limited data sets, but significant deviations from the usual "upside-down parabola" (cf. [Gol96]) and relative changes might be detectable.

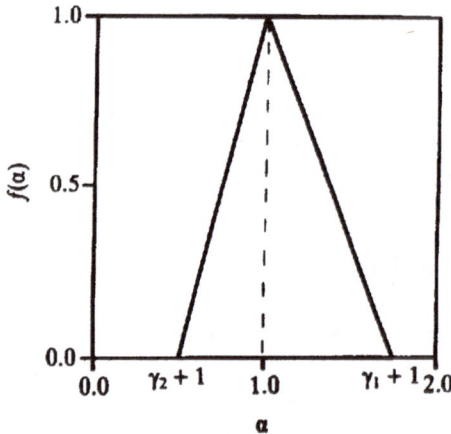

Fig. 6.6. Example of an $f(\alpha)$ spectrum obtained from a singular probability distribution at the critical point of a phase transition (from Beck and Schlögl (1993))

Extensive literature on phase transitions exists (e.g. [BS93] and references therein) and a treatment in the context of earthquakes and self-organised

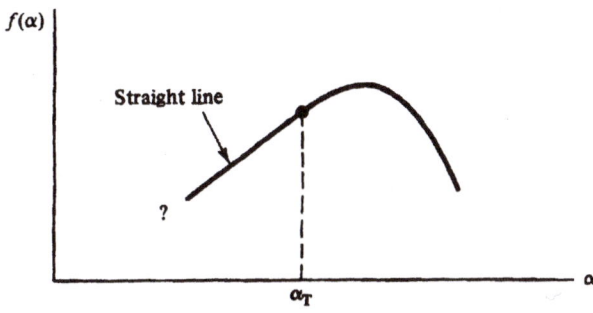

Fig. 6.7. $f(\alpha)$ spectrum obtained for the Henon map. α_T denotes the phase transition point (from Ott (1993))

criticality may be found in Hooge *et al.* (1994). In the latter paper (see also [Hoo93]) it is also stated that the seismic process satisfies the requirements for a first order phase transition. While this has implications for the predictability of earthquakes, the issue of predictability if addressed in Part 2 of this work— here only empirical evidence for multifractal precursors is sought.

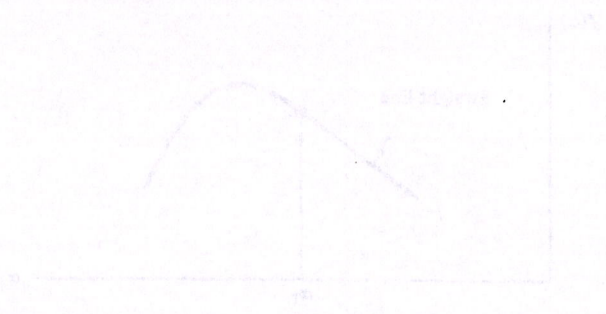

7. Fractal Precursory Behaviour

7.1 The Data

(Multi)-fractal observations seem promising for the detection of anomalous seismicity or even precursory behaviour. In this chapter, most of the fractal concepts introduced earlier will be applied to the temporal evolution of seismicity prior to the m = 6.9 Hyogo-ken Nanbu (Kobe) earthquake which occurred on Jan. 17, 1995 at 34.601° lat. N, 135.032° lon. E near Kobe, Japan (see e.g. [EER95, EQE95] for seismological and other details and Fig. 8.1).

The data used in this analysis was produced by Abuyama Observatory of Disaster Prevention Research Institute, Kyoto University and kindly made available by Dr. Hiroshi Katao. The obtained catalogue started in January 1976 and reached up to the main shock of interest here. During that time no major change in the seismometric network occurred, i.e. sensitivity can be regarded to be constant. The lower magnitude cutoff for completeness of the record is about 1.5 ([Kat95]).

From the original data, a region was selected according to the following criteria:

- Be more or less centred around the epicentre of the main shock
- Contain as many data points as possible
- Do not cut through apparent seismogenic structures at the borders
- Have a size so as not to include too many, possibly unrelated, different seismogenic zones
- Have a size so as not to be too localised, i.e. sparse
- Do not contain a high density of events at the boundaries
- Be roughly square

The conditions given above partially reflect geophysical considerations such as not to include zones of different seismicity and partially show the constraints imposed by numerical considerations. Following the above rules, the area of 33.7° lat. N to 35.7° lat. N and 134.3° lon. E to 136.3° lon. E was selected for analysis. After elimination of events of magnitude < 1.5 and undetermined events, a catalogue of less than 10 000 events resulted. Therefore the lower completeness level was ignored in favour of more data points and the inclusion of microseismicity (as suggested in [HIY92]). Due to the above mentioned continuity of the network, one may assume that the missing events are not concentrated in a certain period, thus having little or no influence on the relative changes of multifractal properties. The minimum detected magnitude was 0.1 and the thus resulting data comprised 26 472 events. The obtained epicentre distribution is shown in Fig. 7.1.

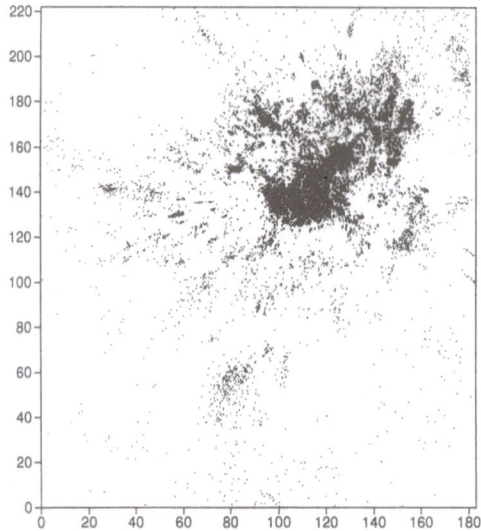

Fig. 7.1. Epicentral distribution of the catalogue selected for the search of possible precursory fractal behaviour

7.2 Overall Properties

The overall fractal properties of the earthquake fields were determined in analogy to the analysis of the landslide fields in Chapter 3 before looking at their temporal variation.

As mentioned in [Gol96], the extreme nonlinear variability of seismic energy and seismic moment poses severe problems for statistical analyses like the one employed to determine the multifractal spectra. The data possesses no central tendency any more which is required for the existence of statistical moments and together with the sparsity of the data makes these fields inappropriate for analysis. Section 5.3 furthermore shows that the errors resulting after the conversion of magnitude to seismic moment or energy are of far too big order. To allow an analysis of a measure describing earthquake "size" in addition to probability of occurrence (in analogy to landslide size and occurrence probability in Chapter 3) however, another measure may be employed.

Following Turcotte (1992), the rupture area A associated with each earthquake was determined from the magnitudes m as follows. First the seismic moment M was determined from

$$\log M = cm + d$$

where c was taken to be 1.5 (cf. [KA75]) and $d = 9.1$ (cf. [HK79, Kan78]) (the total energy in the seismic waves generated by the earthquake may be obtained for $c = 1.44$ and $d = 5.24$). Then A was obtained from

$$M = \alpha A^{3/2}$$

(cf. [KA75]) where α was assumed to be 3.27×10^6 (cf. [Tur92]).

Aki (1981) has used the above equations in connection with the Gutenberg-Richter law

$$\log N(m) = -bm + a \qquad (7.1)$$

where $N(m)$ is the number of earthquakes with (surface) magnitude $m = m \pm 0.1m$ and a and b are constants, to establish a theoretical relation between the well-known "b-value" (the slope in the $\log N(m)$ versus m plot) and the fractal dimension D of the epicentre distribution. He obtained the simple relation

$$D = 2b. \qquad (7.2)$$

The relation seems to hold approximately at first sight because the b-value varies between 0.8 and 1.2 for local as well as global areas (e.g. [Eve70]) and the (capacity) dimension of epicentre distributions varies roughly between 1.6 and almost 2. A direct empirical verification of Eq. (7.2) failed in the case of seismicity in the Tohoku region, however, when Hirata (1989) found a negative correlation between b and D. See also [Hoo93] for a criticism regarding the frequency-magnitude relation (7.1) as a scaling relation. More important is the fact that neither D alone, nor the b-value have shown consistent precursory quality which is why the b-value is not regarded any further.

Unfortunately, multifractal analysis of the rupture area field (as in the case of Hokkaido landslide sizes in Chapter 3) still proved impossible due to very poor scaling behaviour. It is not clear if this was caused by too large an error in the data or by the limited number of data points (recall also that the accuracy of results does not depend on the number of data points alone, but also on the nature of the data itself; cf. the comparison between deterministic and random fractals in Chapter 3). The rupture area field is shown nevertheless in Fig. 7.2 together with the probability of occurrence field for comparison. Also the converted measure for earthquake size was used in a one dimensional analysis and an isotropy analysis below.

It should be pointed out that, when regarding the rupture size of earthquakes, the events cannot be assumed to be point processes anymore. Indeed the events become intrinsically anisotropic in that they have a certain linear extent in a certain direction. If, as in Fig. 7.2, the rupture area field is regarded below a certain cell size, the representation becomes unphysical at least for those events or cells which support a cumulative rupture size greater than the area of the cell. For Fig. 7.2, this limit is 9 km^2 which is in fact surpassed in several cases (the maximum unnormalised cumulative rupture area was 21.10 km^2). It is not immediately obvious how to deal with this problem in the case of a multifractal analysis and Hooge *et al.* (1994) simply defer the problem by saying that only very few events have rupture areas larger than their highest resolution cell size of 2 km^2 (they announce, however, future research involving Lie analysis which would include tensorial information).

Figure 7.2 prominently displays the very different overall behaviours of earthquake location and size for the selected area. The big dominant cluster towards the NE of the area (cf. Fig. 7.1) possesses the highest probabilities for the occurrence of earthquakes (the highest peak represents 704 earthquakes in 9 km^2 when not normalised) while the large events are scattered over the whole area. The area of the dominant cluster might thus be regarded to continuously release seismic energy at a small scale while large events happen intermittently in the whole area. The temporal evolution that led to the fields regarded here is the object of research below. From the lower part of Fig. 7.2 it also becomes clear that even the smoother measure of earthquake rupture size (smooth in comparison to seismic energy or moment) is dominated by a few peaks. This is quite different from the properties of landslide fields as shown in Fig. 3.2 in Chapter 3. The latter might well explain the inapplicability of the multifractal analysis method employed here.

7.2.1 Multifractal Spectrum

As summarised in Chapter 3, the list of publications concerned with the multifractal analysis of earthquake catalogues (mainly the two-dimensional epicentre distribution), is rather short:

Geilikman *et.al* (1990) point out that it is not sufficient to characterise the purely geometric properties of the spatial distribution of seismicity only (monofractal approach), but that the probabilities with which any given parts are visited must also be taken into account (multifractal approach). The authors analyse seismicity in the three seismic regions Pamir-Tyan Shan, Caucasus and California, considering catalogues of as little as 1651 up to 8963 events. The sizes of the areas were 100×100 km or smaller. Shown finally are the $f(\alpha) - \alpha$ curves and an interpretation according to the most and least intensive clustering and the number of such clusters is given. α_{min} is found to be about 1.1 on average, $\overline{\alpha_{max}}$ lies at 4.2, the respective extremal values of f are 0.07 and 0.21. The regional variation in $f(\alpha_{min})$ and $f(\alpha_{max})$ is much greater than the variation in the extremal local scaling exponents. While α_0 tends to be about 2.0 for all regions, the overall shape of the multifractal spectra thus varies considerably from region to region. The conclusions of Geilikman *et.al* (1990) are that seismicity may and must be characterised by multifractal analysis and that further investigation into the use within earthquake prediction must be carried out. The difference of the multifractal spectra from region to region is interesting to note because it does not agree well with the idea of earthquakes as a universal multifractal (cf. [Gol96]) as

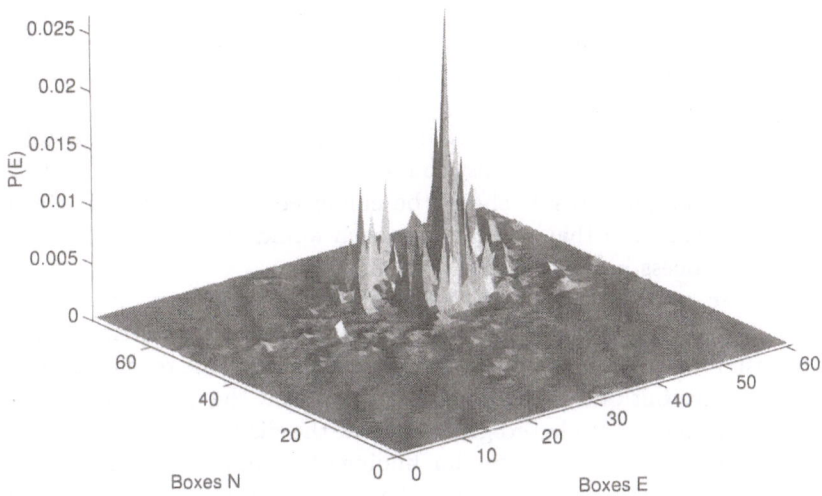

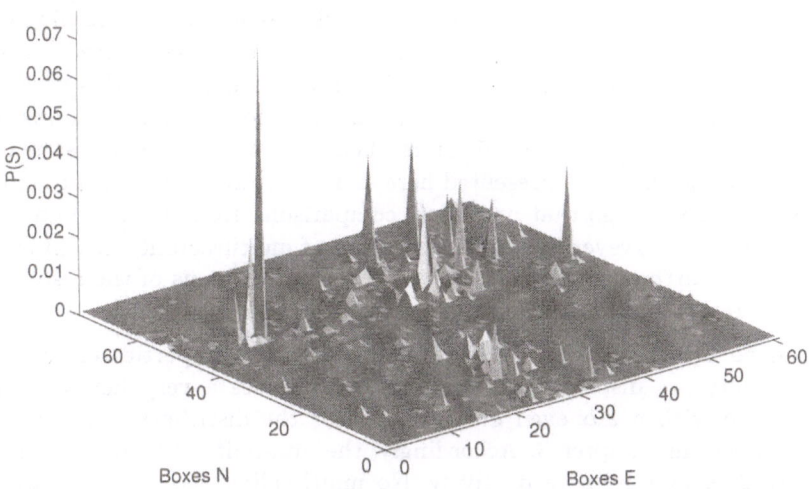

Fig. 7.2. Perspective three-dimensional views of the epicentre distribution (above) and the earthquake size distribution (below) in the area selected for analysis of possible precursory fractal behaviour. The fields were generated by calculating the normalised cumulative probabilities and sizes for 3 km by 3 km cells and thus represent the respective measures at that resolution

put forward by Hooge *et al.* (1994). Recall that universality of the earthquake process would mean independence of the seismic process of many of the geological details (cf. also [Fed88]). In fact, universality would probably imply that observation of spatial and temporal clustering of earthquakes is not useful for earthquake prediction.

Hirata and Imoto (1991) did a multifractal analysis of the hypocentre distribution of micro-earthquakes in the Kanto region of Japan. They analysed a data set of 7013 events of $m >= 2$ in a region of approximately $180 \times 220 \times 60$ km in determining D_q for $q > 0$. D_∞ was estimated to be about 1.7 while D_1, the information dimension, was about 2.18, thus confirming the heterogeneous character of the fractal distribution. The authors conclude that their preliminary incomplete result should be enhanced by determination of the $f(\alpha) - \alpha$ spectrum and that such an approach would foster understanding of the fracture process in heterogeneous media.

The paper by Hirabayashi et al. (1992) [HIY92] has been mentioned in several places already because it not only analyses epicentre distributions but also the distribution of hypocentres and seismic energy. The results are not to be repeated here but one of the main findings is that the distribution of seismic energy is much more heterogeneous than the distribution of occurrence probability (cf. Chapter 3 for similar findings for landslides). Furthermore, that paper is the only one considering the temporal variation of multifractal properties of the spatial distribution of seismicity (cf. Section 6).

Finally, Hooge (1993) and Hooge et.al (1994) take a more theoretical approach to multifractality of earthquakes and present ideas concerning multifractal phase transitions (cf. Section 6) which supposedly lead to self-organised criticality and (multi-)fractality in earthquakes. In this case, not only occurrence probabilities, but their generalisations (one of which is seismic energy) are analysed. Formalism, method and presentation of results does not follow the formalisms presented here and generally employed throughout all the other sources so that numerical comparisons are difficult. Main point of interest here, however, is the confirmation of multifractality, i.e. intermittency of the earthquake process. More detailed discussions of the results are given wherever applicable.

Epicentre Distribution. The overall multifractal properties of the data selected here are displayed in Fig. 7.3. One notices a very heterogeneous multifractal with a $\Delta\alpha$ even greater than for the distribution of landslide sizes as given in Chapter 3. Accordingly, the non-uniformity factors Δ and Δ' were 2.09 and 1.24, respectively. No multiscaling was apparent during determination of the slopes, the scaling region extending from 13.87 km to 178.84 km, i.e. almost up to the extent of the observation area. Following the line of argument in [Gol96], one may say that the present result once more confirms the multifractality of the earthquake phenomenon.

Temporal Distribution. Only Godano and Caruso (1995) have analysed the multifractal properties of the temporal distribution of earthquakes (not to be confused with the analysis of the temporal behaviour of spatial distribution—there, the only reference is [HIY92]). These authors confirm the multifractality of temporal clustering of earthquakes and point out that

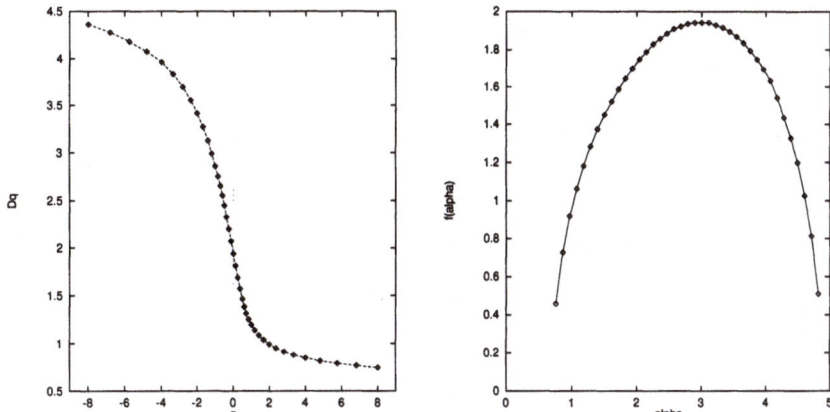

Fig. 7.3. Spectrum of generalised dimensions and multifractal $f(\alpha)-\alpha$ curve for the epicentre distribution of the catalogue selected for analysis of possible precursory behaviour

this implies that the clustering cannot be regarded to be constant, thus making it impossible to model the distribution of earthquakes by a Poissonian process. For earthquakes on a Hawaiian island and in five different regions in Italy, the above authors find D_q curves with a varying degree of inhomogeneity: the maximum $D_{-\infty}$ is about 1.5, the minimum D_{∞} is about 0.1. The authors conclude that a continuous monitoring of multifractal properties could be a major enhancement to earthquake prediction but conduct no such research.

In the multifractal analysis of the temporal distribution of earthquakes in the selected area, a one dimensional data set of cumulative times of occurrences was generated first. Such a data set is analogous to a Cantor set and requires much less calculation time than a field and also yields more accurate results. The results are displayed in Fig. 7.4. The log − log plots were very convincing in that they produced linear fits with a very small mean square error over a very wide range of cumulative time (the whole range was 0 to $7.421\ldots \times 10^{14}$ secs, the scaling region went from 2.264×10^8 to 1.855×10^{12} secs). The non-uniformity factors were $\Delta = 0.53 \pm 0.01$ and $\Delta' = 0.13 \pm 0.008$. Interestingly, the D_q curve has a very gentle slope for $q < 0$, i.e. saturates quickly towards $D_{-\infty}$, while the convergence to D_{∞} is very steep. The latter implies that the sparsely populated temporal vicinities are less heterogeneous than the intervals in which many earthquakes occur. Thus results the leaning $f(\alpha) - \alpha$ curve with a large $f(\alpha_{max})$. Such a multifractal spectrum indicates that there are many more sparsely populated intervals within the multifractal than there are densely populated ones. It will be interesting to observe the temporal fluctuation of this ratio during the evolution of seismicity. Note that $f(\alpha_0)$ is very close to 1, indicating that in general there are no times were no earthquakes can occur.

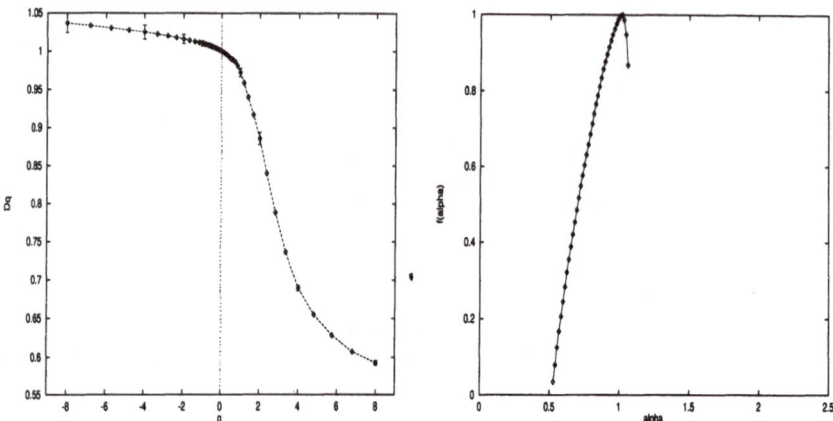

Fig. 7.4. Spectrum of generalised dimensions and multifractal $f(\alpha) - \alpha$ curve for the temporal distribution of earthquakes selected for analysis of possible precursory behaviour

7.2.2 HOT Analysis

Fig. 7.5 shows the results of a Hurst Oriented transform and the consequent Hurst exponents as a function of direction. One notices the rather "spiky" form of the rose plot which is the result of the sparsity of the data (this effect prevails in spite of the neglection of differences between occupied and void cells or better, results from this technique—but otherwise no meaningful results would be obtained at all). Despite the above mentioned drawback, the orientation of the major axis was found to be 73.45° and the ratio of the axes was 0.60, indicating a certain degree of anisotropy of the fractal dimension of the epicentre distribution. Indeed the fractal dimension varied between about 1.99 to about 1.67 with a mean value of 1.84±0.12. Theoretically, the latter value should agree with the result for D_0 of e.g. a box-counting analysis and with $f(\alpha_0)$. Comparing Fig. 7.3, one finds such a rough agreement but, as pointed out earlier, Hurst analysis yields no accurate absolute results for D. Looking at Fig. 7.6 where the results for the earthquake rupture size field are given reveals the inapplicability of the HOT transform for sparse intermittent data: especially the HOT table shows that the determined ranges are dominated by a few comparatively very large values; discontinuities in the individual Hurst plots for each row (i.e. direction) are therefore to be expected. Thus results the even more singular rose plot in Fig. 7.6 which must

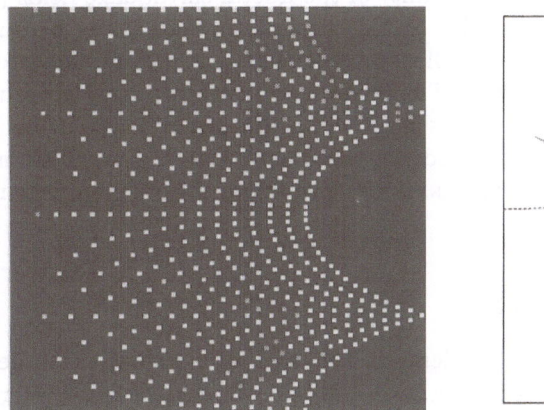 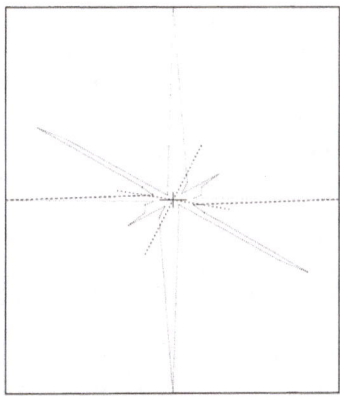

Fig. 7.5. HOT table and rose plot of H for the distribution of epicentres

be interpreted with care. The highest fractal dimension of about 1.79 occurs in a direction of 46.38°, the lowest value of D is roughly 0.93. The average fractal dimension is 1.33±0.35, the degree of anisotropy is 0.5. Again, comparison between data sets of different overall characteristics are not meaningful but one might conclude that the size field is more anisotropic than the epicentre distribution field. Also it possesses generally a lower fractal dimension. A comparison of these results with the anisotropic properties of an aftershock sequence will be carried out in Section 8.3.

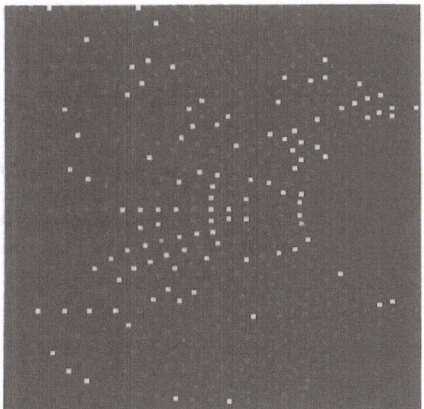 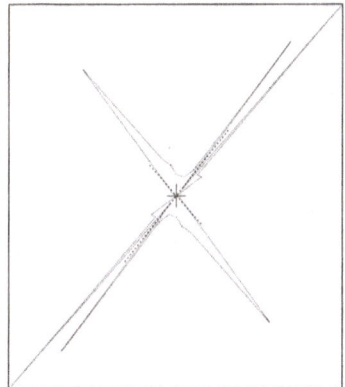

Fig. 7.6. HOT table and rose plot of H for the distribution of rupture areas

A moving analysis of the directionality of the fractal earthquake fields (at least the epicentre distribution field) would certainly be interesting because the distribution of epicentres as well as earthquake size reflects the stress fields in the crust (cf. also [OU86]). The latter fields might undergo some re-orientation or re-distribution in the seismic cycle which might well show up in the directionality of the fractal fields regarded here. Due to computational limitations however, such an analysis has been postponed to consequent research.

7.2.3 Configuration Entropy

The configuration entropy, its implementation and its application to geophysical fields as well as its interpretation has been discussed in Chapter 3 and is applied here in a similar way. An entropy analysis of the data considered here yielded the curve shown in Fig. 7.7. One notices a typical curve as expected for a multifractal of low lacunarity. The latter finding explains why the $\log - \log$ plots used for the determination of D_q and $f(\alpha)$ showed smooth linear behaviour without any steps. The low value of r^* at about 810 m will be of interest when comparing the respective corresponding values in the moving analysis. As pointed out earlier, the value of H^* is only of interest when comparing data sets of the same overall characteristics (i.e. same spatial extent and number of data points) and will gain significance in the analysis of temporal behaviour below.

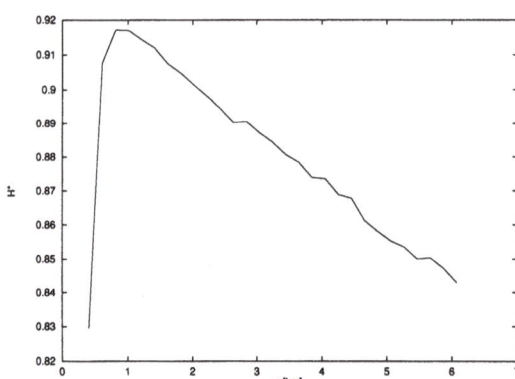

Fig. 7.7. Configuration Entropy curve for the epicentre distribution selected for analysis of possible precursory behaviour

7.3 Temporal Variation of Properties

To detect possible precursory fractal behaviour, a record prior to the main shock as long as possible but also as quiet as possible is desirable. Not to miss the possible influence of large earthquakes that occurred in the vicinity of the selected area mentioned above, the earthquake history was recorded for the greater area 33.0° lat. N to 36.0° lat. N and 134.0° lon. E to 137.0° lon. E. Fig. 7.8 is a magnitude versus time plot for all events with $m >= 3$ (726 events) from Jan. 1, 1976 up to immediately before the Hyogo-ken Nanbu earthquake which reveals that several earthquakes of a magnitude m of about 5 occurred in the interval. The largest earthquake in the greater area besides the Jan. 17 mainshock had a magnitude of 5.6 and occurred on May 30, 1984. Table 7.1 gives a listing of all earthquakes with $m >= 4.5$ contained in the greater area including the respective event and day number with respect to the selected catalogue[1].

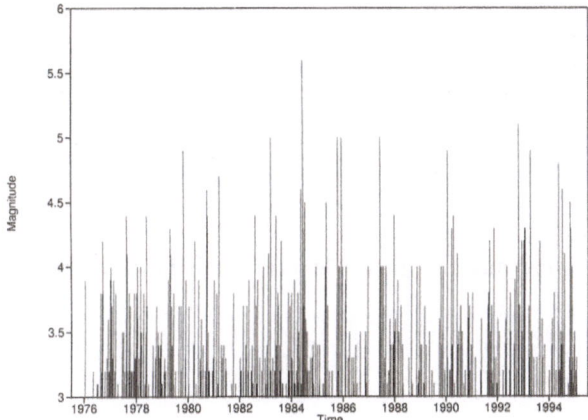

Fig. 7.8. Magnitude ($m >= 3.0$) versus time diagram for the selected data set

    Several of the large events occur clustered due to foreshock as well as aftershock activity. Thus, one would not expect individual precursors but rather precursory behaviour for each of the "cycles"[2].

    To further simplify the correlation between temporal behaviour of fractal parameters and earthquake history in the selected area, Fig. 7.9 repeats Fig. 7.8 in that it also shows magnitude but this time plots it versus event number. The lower part gives the daily frequency of earthquakes together with the temporal location of earthquakes of $m >= 4.5$ (diamonds) and $m >= 5.0$ (crosses). The latter figure will be convenient for the discussion of the temporal behaviour of fractal parameters below. Note that Fig. 7.9 shows the activity within the smaller *selected* area (except for the individu-

---

[1] If the event fell outside the area selected for numerical analysis, the numbers correspond to the closest event contained. This will simplify later correlation between temporal behaviour and event.

[2] Cycle not in the sense of a spectral peak but in the sense of times of increased activity, i.e. intermittent behaviour.

Table 7.1. Listing of earthquakes of $m \geq 4.5$ in the area selected for analysis of possible precursory fractal behaviour

Yr	Mo	Dd	Dd#	m	#	Yr	Mo	Dd	Dd#	m	#
76	7	26	208	4.8	701	85	9	20	3551	4.8	14330
76	11	11	316	4.6	1203	85	10	3	3564	5.0	14373
77	3	5	430	4.7	1650	85	11	27	3619	5.0	14841
77	8	6	583	4.6	2314	87	5	28	4166	5.0	17395
78	6	20	901	4.6	3771	88	1	25	4408	4.6	19786
79	10	16	1385	4.9	6185	89	2	19	4799	5.0	20099
80	8	5	1679	4.6	7302	90	1	11	5125	4.9	21136
80	9	11	1716	4.6	7440	90	9	29	5386	5.4	23319
81	3	6	1892	4.7	8167	92	7	30	6056	5.3	23945
83	1	26	2583	4.8	10557	92	10	17	6135	5.1	24139
83	3	6	2622	5.0	10741	93·	3	14	6283	5.1	24436
84	5	5	3047	4.6	12309	93	4	2	6302	4.9	24480
84	5	30	3073	5.6	12470	94	5	8	6703	4.8	25358
84	5	30	3073	4.9	12472	94	5	22	6717	4.9	25395
84	5	30	3073	5.0	12474	94	5	28	6723	5.2	25408
84	6	25	3099	4.5	12635	94	6	28	6754	4.6	25507
84	8	9	3144	4.8	12859	94	10	16	6864	4.5	26046
85	4	27	3404	4.5	13762	94	12	23	6932	4.6	26423
						95	1	17	last	6.9	last

ally indicated events above a certain magnitude level) as this is what will be numerically analysed.

7.3.1 Temporal Multifractal

Because of the remaining uncertainties in the determination of multifractal spectra which require manual intervention mainly during evaluation of the slopes of the $\log - \log$ plots (this is despite the ideas for automatic simultaneous detection of the scaling region as detailed in Section 3.4), a fully automatic procedure is questionable. A moving analysis on the other hand requires such an amount of spectra to be determined that a manual approach is not feasible either. In the analyses carried out in the next two sections, D_q for $q < -2$ and thus the values of α_{max} and $f(\alpha_{max})$ must be regarded carefully. Δ' was obtained in addition to Δ for the same reason.

Due to previous experiences with small data sets, the window size was increased to 5000 events as compared to the 500 events used by Hirabayashi et al. (1992). The overlap was reduced to 2% (100 events) as compared to 50% to enhance the temporal resolution of the resulting time series. The values determined for each window thus present integral fractal properties over a time of about 3 years with a Δt of roughly 25 days on average. The obtained values are plotted versus the event number and centred on the centre of each respective interval (e.g., the first point of any of the following curves, obtained by analysing events no. 1 to no. 5000 is plotted at event no. 2500). For the latter reason alone, one cannot expect exactly corresponding

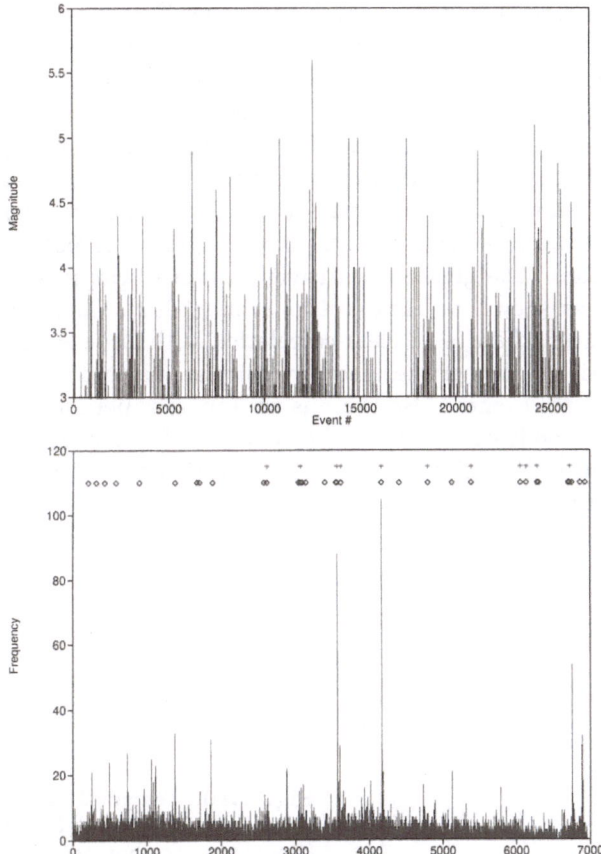

Fig. 7.9. Magnitude vs. event number (above) and daily frequency of earthquakes (below) with earthquakes of $m >= 4.5$ (diamonds) and $m >= 5.0$ (crosses) indicated for the area selected for analysis of precursory fractal behaviour

locations of e.g. peaks in D_q and in seismic activity as shown in Fig. 7.9. Rather, the interesting feature to look for will be (precursory) trends. Other combinations of window lengths and overlap might certainly be interesting in future analyses.

Epicentre Distribution. Remember that changes in α_{min} and $f(\alpha_{min})$ reflect changes in areas of high seismic activity in that an increase of α_{min} means that the most populated vicinities at a given time get more populated (the reverse is true for a decrease in α_{min}) and that an increase in $f(\alpha_{min})$ means an increase in the number of such areas (with local scaling α_{min}) while a decrease indicates that dense areas get less in number. An analogous interpretation applies to α_{max} and $f(\alpha_{max})$, only that the least densely populated vicinities (areas of lowest seismic activity) are characterised.

The outcome of the moving analysis of the distribution of epicentres is shown in Figs. 7.10 to 7.13:

Fig. 7.10 shows the D_q curves for $q = -2, 0, 1, 2$ (from top to bottom), i.e. the generalised dimension recommended for observation by Hirabayashi *et al.* (1992), the capacity dimension, the information dimension and finally the correlation dimension. This figure is shown first despite its lack of physical interpretation because the consequent time series are derived from this data (cf. [Gol96]). At first glance it is confirmed that the capacity dimension D_0 shows very little sensitivity to the evolution of the seismicity pattern. Also D_1 and D_2 are rather smooth and show no apparent trends. With increasing order q however, variation increases as more and more emphasis is put on densely populated subregions. The greatest sensitivity, however, can indeed be observed for D_{-2} as was shown by Hirabayashi *et al.* (1992). Contrary to the latter authors, however, there is no clear correlation with large events or periods of enhanced activity. Rather there are several plateaus and spikes not clearly correlated with big events. Also there is no apparent persistence in the time series. It thus confirmes what has been said in Section 6: the observation of a single generalised dimension is not a particularly suitable method to detect precursory behaviour.

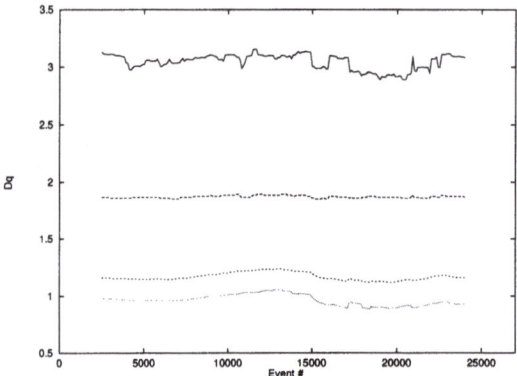

Fig. 7.10. Temporal variation of D_q for $q = -2, 0, 1, 2$ (from top to bottom) for the epicentre distribution selected for analysis of possible fractal precursory behaviour

Fig. 7.11 is similar to Fig. 7.10 in that it gives α_{min} (solid) and α_{max} (dashed), i.e. D_∞ and $D_{-\infty}$ as obtained from the horizontal saturation regimes of the individual D_q curves. α_{min} has been offset by $+3.4$ for clarity. Also indicated are the two largest events besides the final $m = 6.9$ earthquake: the $m = 5.6$ event no. 12 470 on May 30, 1984 and the $m = 5.4$ earthquake no. 23 319 on Sept. 29, 1990 (diamonds from the left). These values of α have a physical meaning in that they directly describe the scaling of the densest and sparsest earthquake clusters encountered throughout the whole catalogue. If α_{min} increases, it means that the densest clusters get less dense at a given time, if α_{min} decreases the densest of all clusters get even more populated. As for α_{max}, an increase (i.e. shift to the right) indicates that seismicity in the most inactive region decreases even more; if α_{max} decreases, the sparsest clusters get more active. α_{max} is found to possess several abrupt steps of great magnitude after which the curve returns to a slow trend quite similar to α_{min}. Keeping the notorious unreliability of α_{max} in mind, it is not surprising that some of these peaks do not correspond to the earthquake history as depicted in Fig. 7.9 and must probably be attributed to numerical error. Note however, the increasing trends before the two major earthquakes indicated: both events take place at about the maxima of α_{max}. This would mean, if generally true, that seismicity in the sparsest regions gets sparser and sparser until a major event occurs and might agree with e.g. Haikun's (1993) observation of "energy concentration". After the event, seismicity in the sparsest regions increases rather rapidly.

Regarding the numerically more reliable α_{min}, one learns about the evolution of the areas of strongest seismic activity. The curve lacks the offset plateaus of α_{max} and, as already mentioned, roughly follows the trend of α_{max}. Although the second maximum is not as pronounced as in α_{max}, both major earthquakes also happen at the maxima. This behaviour would imply that seismic activity in the most active regions also goes down before a main shock and recovers afterwards. As D_0 is rather constant, one might conclude that the level of overall seismicity decreases before a big earthquake happens. The latter would not agree with the idea of energy concentration.

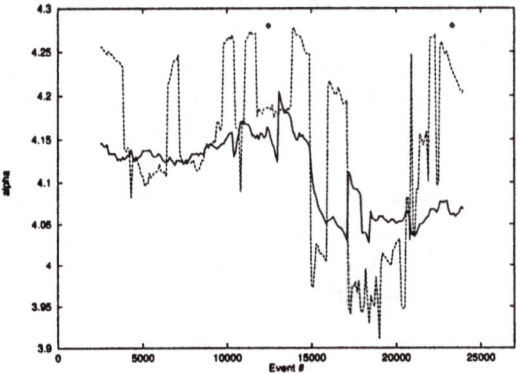

Fig. 7.11. Temporal variation of α_{min} (solid) and α_{max} (dashed) for the epicentre distribution selected for analysis of possible fractal precursory behaviour

Further information is gained from Fig. 7.12, where $f(\alpha_{min})$ (solid) and $f(\alpha_{max})$ (dashed)are shown. These curves show how many of the respective subregions of densest and sparsest clustering, respectively, exist at every time. Again the curve associated with α_{max} is found to fluctuate violently in a seemingly random fashion. If at all, a decreasing trend before the two strongest events may be observed. It would mean that the number of extremely sparse clusters decreases before main shocks (while the sparsity of these clusters increases simultaneously). Possibly more interesting and significant is the behaviour of $f(\alpha_{min})$: a very clear minimum occurs during the $m = 5.6$ event. The trend towards this minimum commences several months before the event. A decrease in $f(\alpha_{min})$ signifies that the number of extremely active clusters of seismic activity goes down. After the large event no. 12 470, however, the number of highly active clusters assumes an almost constant level which does not change prior or during the second largest $m = 5.4$ event.

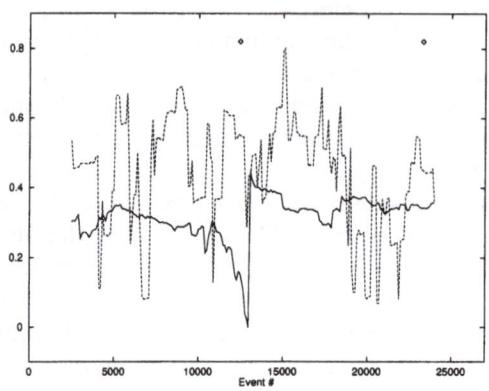

Fig. 7.12. Temporal variation of $f(\alpha_{min})$ (solid) and $f(\alpha_{max})$ (dashed) for the epicentre distribution selected for analysis of possible fractal precursory behaviour

The variation of $\Delta\alpha$ is summarised in Fig. 7.13, where the non-uniformity factors Δ (solid) and Δ' (dashed) are displayed. Δ' has been offset by $+0.5$ for clarity. It becomes obvious that the non-uniformity factor is dominated here by α_{min}. The normalisation by α_0 has almost no effect due to the small variability of D_0. In this case, Δ and also the numerically more reliable Δ' yield no additional information at all. Except for the final increase in heterogeneity before the $m = 5.4$ event, the signal becomes less meaningful as for the occurrence of strong earthquakes. Also because the variation in f is not included in Δ, the separate observation of the parameters discussed above seems a better approach for monitoring seismicity. Appendix B suggests how to possibly enhance the precursory quality of the above multifractal parameters by a fuzzy cluster analysis prior to multifractal analysis.

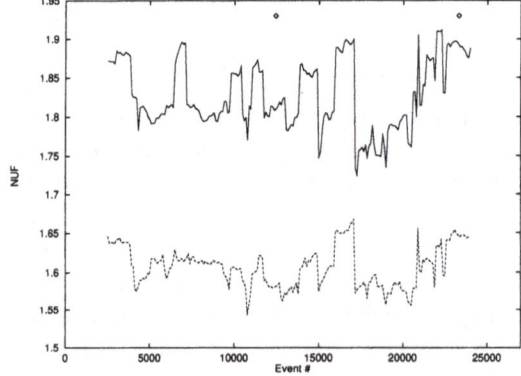

Fig. 7.13. Temporal variation of Δ (solid) and Δ' (dashed) for the epicentre distribution selected for analysis of possible fractal precursory behaviour

Temporal Distribution. The multifractal variation of the distribution of earthquakes in time is shown in Fig. 7.14 in a fashion similar to Fig. 7.10. Plotted are D_{-2}, D_0, D_1 and D_2 from top to bottom. Contrary to the distribution in space (see Fig. 7.10), D_q for $q > 0$ is most sensitive now. In fact, D_{-2} shows almost no variation at all. The strong variability of D_2 indicates that variation in the multifractal spectrum originates from changes within the strongly clustered intervals—strongly clustered intervals get more or less populated while sparsely populated intervals stay at the same low level of seismicity throughout the whole observation time of about 18 years.

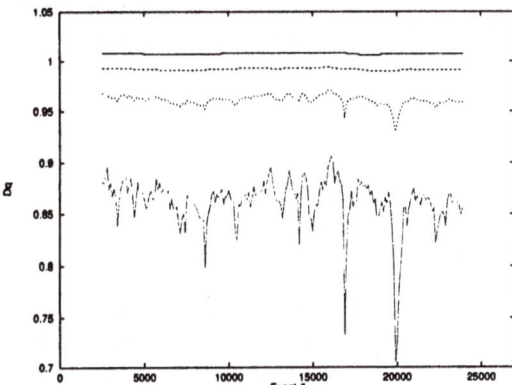

Fig. 7.14. Temporal variation of D_q for $q = -2, 0, 1, 2$ (from top to bottom) for the temporal distribution of earthquakes selected for analysis of possible fractal precursory behaviour

Accordingly, Fig. 7.15 only shows α_{min} (top) and $f(\alpha_{min})$ (bottom) (there was no significant variation in $f(\alpha_{max})$ either). The variation of α_{min} is similar to the one of D_2, i.e. the correlation dimension. While several of the pronounced peaks correspond to seismic activity, such as the one at the indicated $m = 5.6$ earthquake, no long-term (precursory) behaviour is apparent. The aforementioned peak indicates that the local scaling exponent of the densest temporal clustering increases co-seismically. This was to be expected due to the very rapid succession of aftershocks and confirms the physical meaningfulness of α_{min}. $f(\alpha_{min})$ shows very small sensitivity in numerical range of small scale fluctuations but also in the long-run—the number of extremely dense temporal clusters can be said to be constant throughout the whole observation time.

Finally, Fig. 7.16 gives the non-uniformity factors Δ and Δ' which are merely a negative copy of α_{min} due to the constance of α_{max} and α_0. They demonstrate, however, that, e.g. during the $m = 5.6$ event, heterogeneity of the multifractal temporal distribution increases during a major event.

The temporal variation of the multifractal temporal distribution of earthquakes may be said to be much less significant with respect to the actually observed seismicity than the one of the distribution of epicentres. While the result of a multifractal analysis of one-dimensional data is numerically more reliable, it is not surprising that it also yields less information. As mentioned

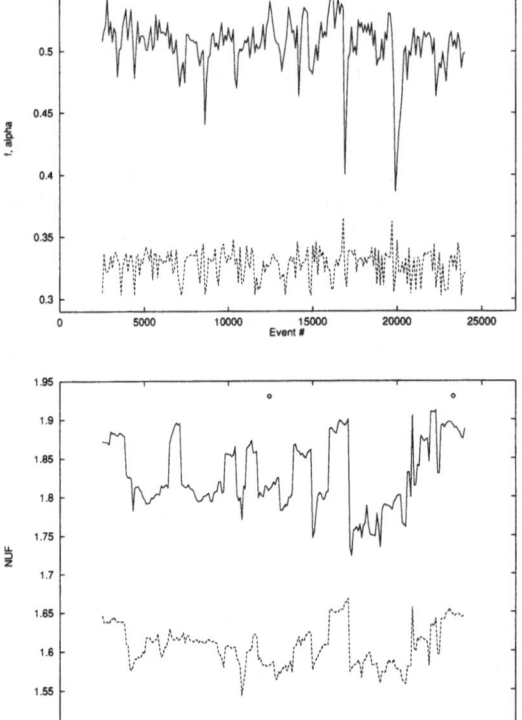

Fig. 7.15. Temporal variation of α_{min} (top) and $f(\alpha_{min})$ (bottom) for the temporal distribution of earthquakes selected for analysis of possible fractal precursory behaviour

Fig. 7.16. Temporal variation of Δ (top) and Δ' (bottom) for the temporal distribution of earthquakes selected for analysis of possible fractal precursory behaviour

earlier, an analysis of the five-dimensional complete earthquake-space would be best from the physical point of view but is numerically not feasible.

7.3.2 Temporal Configuration Entropy

Because data sets of the same number of events are compared during a moving analysis, the value of H^* gains significance in addition to r^*. Thus it was attempted to determine both the maximum of the configuration entropy and the optimum entropy length.

The minimum window size was adopted from Section 7.3.1 although a smaller size might be feasible from the numerical point of view. Setting the overlap to 100 points still produced sufficient temporal resolution and meant 216 independent configuration entropy analyses which took about 12 hours on a Sun workstation (using 2000 randomly selected reference points in every analysis). Thus it becomes clear again that computational limitations unfortunately have a large role to play in this work.

Figure 7.17 gives the temporal variation of r^* (above) and H^* (below) with the two largest earthquakes besides the final event indicated by vertical lines. At first sight it becomes obvious that both measures are sensitive to the evolution of seismicity—the obtained timeseries do not seem to fluctuate randomly but rather follow certain trends and possess a certain dynamics. This fact may be regarded as most encouraging because of the much higher accuracy and reliability of the configuration entropy analysis as compared to a multifractal analysis of small data sets. While $r^*(t)$ shows discontinuities in the form of plateaus, $H^*(t)$ seems to be a differentiable process. An analysis of the dynamical properties of $H^*(t)$ was attempted in Part II of this work but the time series was found to be too short for meaningful analysis. When comparing the above two parameters further, they seem to possess an independent behaviour or rather a negative correlation. Roughly, when $H^*(t)$ undergoes a maximum, $r^*(t)$ has a minimum and vice versa. This would mean that when disorder (or information) is at its maximum, the optimum entropy length, i.e. the resolution at which this disorder is best displayed, is small compared to more ordered seismicity.

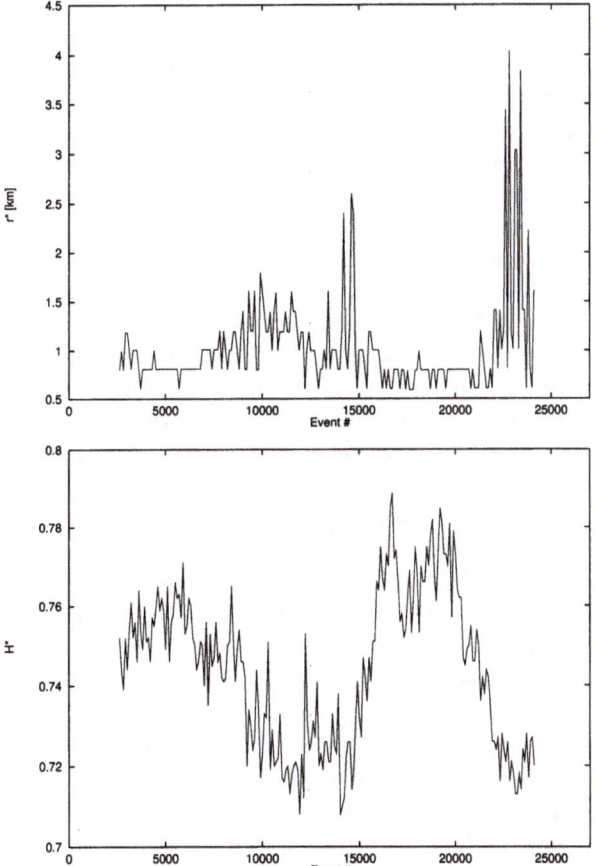

Fig. 7.17. Results of a moving entropy analysis: $r^*(t)$ (above) and $H^*(t)$ (below). Also indicated are the two largest earthquakes besides the final $m = 6.9$ event

$H^*(t)$ has two dominant peaks which might well correspond to the two dominant occurrences of strong seismic activity encountered in the record. The range of fluctuation is numerically rather small and the average value agrees with the result in Section 7.2.3, taking the different number of data points into account (cf. Chapter 3). The first maximum lies at about event number 6000 and then the curve decays almost linearly until about event 12 000. After that, a local maximum follows over roughly 2500 events before the entropy value steeply increases again until about event number 17 400 where the overall maximum of disorder is reached. The subsequent peak is reached after a short decaying period of about 2300 events around event 19 700. Finally, another steep decline occurs until event 23 000 from where the disorder seems to assume an increasing trend again.

As the size of earthquakes has no influence on the analysis, it is not obvious whether the changes are primarily induced by the occurrence of large events (with their associated seismicity of highly clustered aftershocks, see Chapter 8) or the occurrence of earthquake swarms without large events. Recall that the number of events per analysis is kept constant, the spatio-temporal coupling of earthquakes is therefore eliminated by shrinking or expanding the individual intervals in time. When correlating $H^*(t)$ with the daily frequency of earthquakes, as given in Fig. 7.9, it becomes apparent that active times roughly correspond to high disorder while the epicentre distribution of quiet intervals is more ordered. The absolute maximum of $H^*(t)$ corresponds to the maximum daily earthquake frequency associated with the $m = 5.0$ event no. 17 395 on day 4166. One may thus conclude that the earthquakes occurring during highly active periods as expressed by the peaks in the lower part of Fig. 7.9 possess a peculiar entropic behaviour as well. On the other hand, there in no minimum in $H^*(t)$ between the two largest activity peaks in Fig. 7.9 and the final broad peak in $H^*(t)$ approximately centred on event no. 19 700 has no counterpart in earthquake frequency at all. Trying to correlate details of $H^*(t)$ with the event magnitudes as displayed in the upper part of Fig. 7.9 however, gives an even less clear picture: although several spikes of the entropy value coincide with larger earthquakes at the small scale, there is no apparent connection between the cyclic behaviour of $H^*(t)$ at the large scale and the magnitude history, i.e. no extremely large earthquakes occur at these large-scale peaks.

Instead, the cyclic behaviour of $H^*(t)$ might possibly be regarded as precursory for the two main events in the whole interval besides the Hyogo-ken Nanbu earthquake: the first cycle reaches its minimum before the $m = 5.6$ event after decaying for about 4 years, the second cycle reaches its minimum very shortly before the $m = 5.4$ event no. 23 319 on Sept. 29, 1990 after decaying for about 3 years. Note that the latter event does not occur in the magnitude versus time or magnitude versus event number plots in Figs. 7.8 and 7.9 because the earthquake occurred slightly outside of the selected area at 35.001° lat. N, 134.277° lon. E. The fact that its effect seems to show nevertheless indicates that the choice of area for an entropy analysis is not critical—an advantage over several other methods based on analysis of seismicity patterns. Although there are only two examples here, they are also quite convincing as there are no obvious other explanations for the decays in $H^*(t)$. The latter possible precursor might also be associated with the final $m = 6.9$ earthquake but a determination of $H^*(t)$ at finer resolution would be required to judge this. The evolution of the entropy value beyond the end of the current analysis would also shed some light on this. If it is true that $H^*(t)$ decays prior to major earthquakes, it would mean that the degree of disorder decreases   until a major release of seismic energy occurs. After the earthquake, seismicity goes back to a more disordered state until a new cycle begins.

As mentioned above, $r^*(t)$ shows a roughly anti-cyclic behaviour with respect to $H^*(t)$. While there are co-seismic peaks such as the ones at event nos. 14 373 and 14 841 (compare Figs. 7.17 and 7.9), the largest two events mentioned above are not connected with simultaneous peaks in $r^*(t)$. Instead, $r^*(t)$ is still in its first broad peak when the $m = 5.6$ event occurs. The largest peak in Fig. 7.17 again has no counterpart in the magnitude history but occurs immediately before the $m = 5.4$ event. The possible precursory quality of $r^*(t)$ is more questionable than that of $H^*(t)$ although the frequency statistics in Fig. 7.9 gives no explanation for the appearance of the largest peak in $r^*(t)$. Should it be true that r^* increases before major earthquakes, it would mean that the scale (cell size) at which the epicentre distribution displays its maximum information increases during that time while it decreases afterwards.

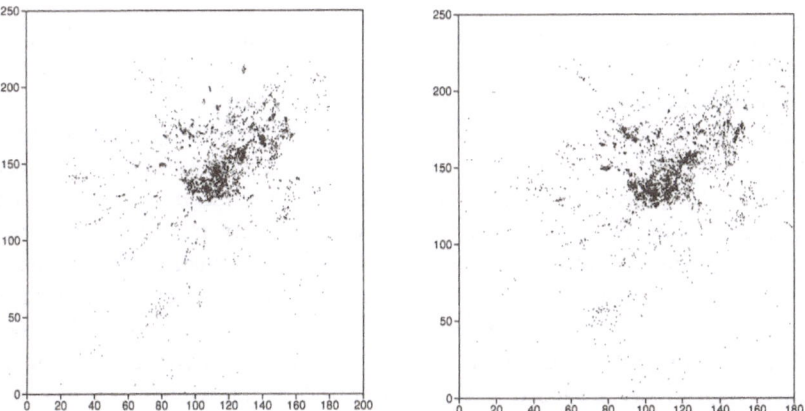

Fig. 7.18. Two example epicentre distributions from which the minimal (left) and maximal (right) values of $H^*(t)$ were obtained

Fig. 7.18 gives two examples of epicentre distributions from which the absolute minimum and absolute maximum respectively of $H^*(t)$ were obtained (shown are the intervals nos. 86 centred on event no. 11 100 and 142 centred on event no. 16700, cf. Fig. 7.17). Analogous distributions for $r^*(t)$ are shown in Fig. 7.19 (intervals nos. 32 centred on event no. 5700 and 203 centred on event no. 22800). The latter figures demonstrate the sensitivity of the configuration entropy for changes in the seismicity pattern which are optically not really discernible.

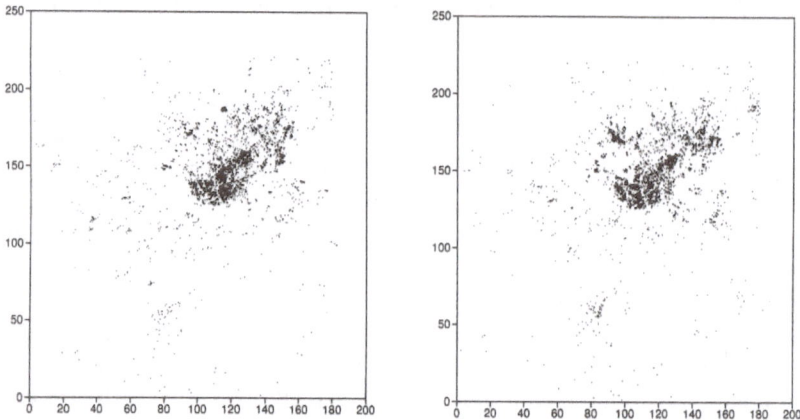

Fig. 7.19. Two example epicentre distributions from which the minimal (left) and maximal (right) values of $r^*(t)$ were obtained

To summarise the results of Chapter 7, it must first be noted that the originally intended goal of finding precursory behaviour for the Hyogo-ken Nanbu earthquake was not achieved. Instead, the temporal behaviour of the multifractal properties of seismicity has shown what might be precursory behaviour for the two other large earthquake events contained in the record. The failure to detect sensitivity to the final event might be attributed to two possible reasons. Firstly, the history of seismicity in the selected region must be regarded to be rather complex, possibly obscuring the signature of the Kobe earthquake. Secondly, the data set might not cover enough time before the latter event to be able to resolve a superposed long-term trend. The first reason could be clarified by conducting the above analyses for a less complex region, i.e. a record with a pronounced quiescence before a major earthquake. Analysis of several other catalogues should definitely be carried out to confirm the shown sensitivity of the concerned fractal parameters to the seismic evolution.

8. Fractal Properties of Aftershocks

Contrary to the search for earthquake precursors, it is of interest to determine the fractal properties of aftershocks as they designate a different stage in the seismic cycle or, in the light of Section 6, another "phase" or "state" of seismicity in a given region.

In the following, several analyses of the aftershocks registered after the Hyogo-ken Nanbu earthquake (cf. Chapter 7) are carried out. The data, registered by the seismic networks of Earthquake Research Institute (ERI), University of Tokyo, Disaster Prevention Research Institute (DPRI), Kyoto University and Faculty of Science, Kochi University, consists of the automatically determined hypocentres (at ERI), times of occurrence and magnitudes and was publicly available[1]. The aforementioned dataset was preferred to a similar one compiled by DPRI[2] because it contained more events and seemed more consistent[3].

Figure 8.1 shows a map of the vicinity of the main shock with epicentres determined by ERI for the time of January 16, 1995, 00:00 JST (i.e. including foreshocks) to January 18, 1995, 18:00 JST. Also shown are known fault-lines (see [Hir89b] for a fractal analysis of fault lines in Japan).

From the raw data, a catalogue consisting of 4903 events was selected in the area of 34.1° lat. N to 35.1° lat. N and 134.6° lon. E to 135.6° lon. E. The resulting data comprised events from Jan. 17, 05:53 to Mar. 31, 06:44 with magnitudes ranging from 0.2 - 4.5 and depths down to 39.9 km. Deeper earthquakes were discarded because of their sparsity (only 46 events with a hypocentral depths > 40 km occurred and must be regarded as faulty anyway) to maintain a relatively densely populated three-dimensional space. For the analyses including the magnitude values, events with unreasonable magnitude (e.g. 9.9 or negative values) were discarded, slightly negative depths were set to zero. Figure 8.2 shows the final data set of epicentres used for analyses, the coordinates transformed into km relative to the SE corner of the selected area.

---

[1] ftp-server `ftp.eri.u-tokyo.ac.jp`
[2] available together with the ERI data
[3] Due to loss of instruments in the main shock, part of the data was gathered using the less accurate spare system.

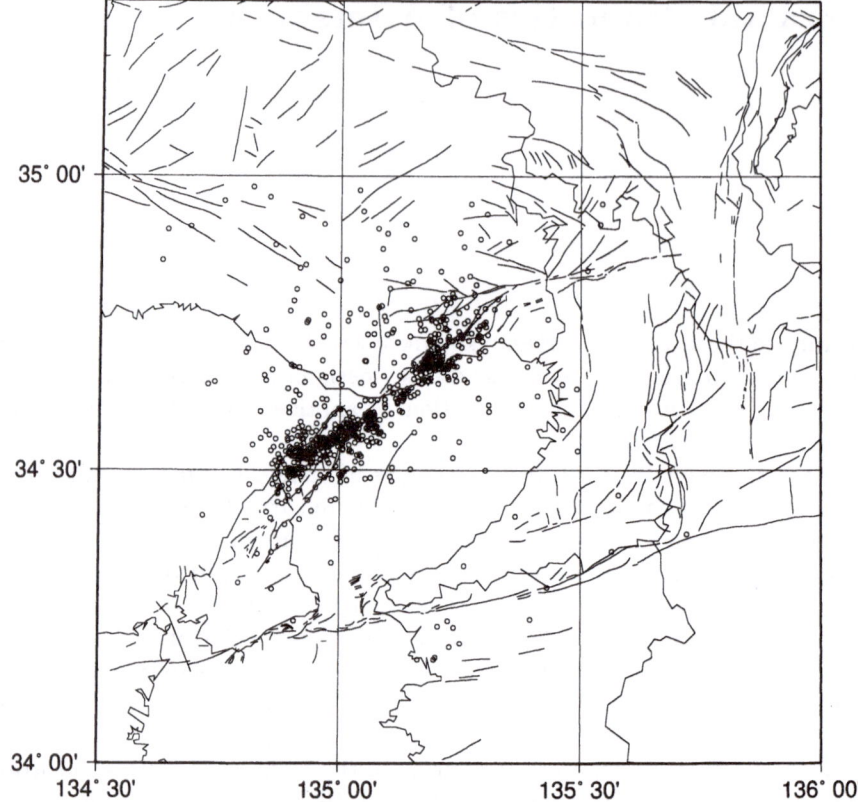

Fig. 8.1. Epicentre distribution in the vicinity of the Hyogo-ken Nanbu earthquake from Jan. 16, 00:00 JST to Jan. 18, 18:00 JST (Earthquake Research Institute, University of Tokyo)

Figure 8.3 gives perspective views of the hypocentre distributions at rotational angles of 30° and 120° respectively to get an impression of the depths distribution as well.

A very preliminary "fractal" analysis is shown in Fig. 8.4, where the number of aftershocks per 12 hours is plotted versus time. One can easily observe the validity of the Omori law mentioned in Chapter 4 , i.e. the scaling behaviour of the temporal distribution of aftershocks. Further analysis of the temporal scaling properties is carried out below. Figure 8.4 together with e.g. fig. 8.3 is a good example of the spatio-temporal coupling of earthquakes as already addressed in Chapter 6.

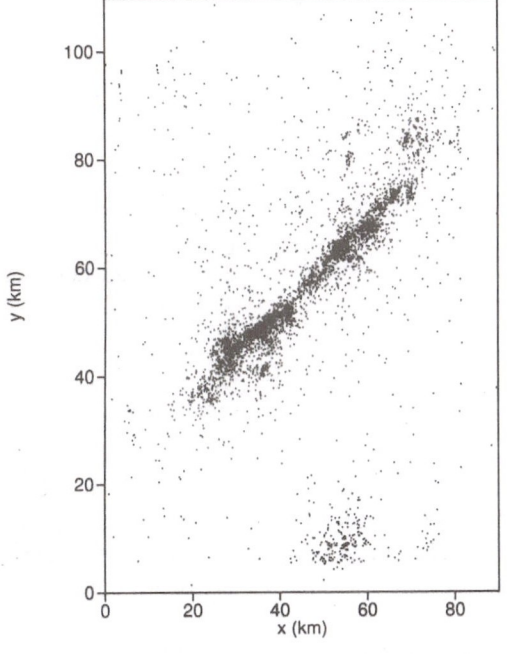

Fig. 8.2. Epicentral after-shock distribution of the Hyogo-ken Nanbu earth-quake as used for analysis

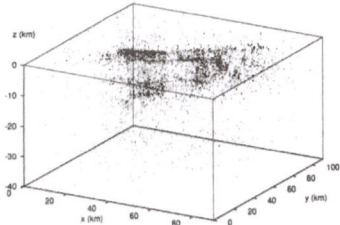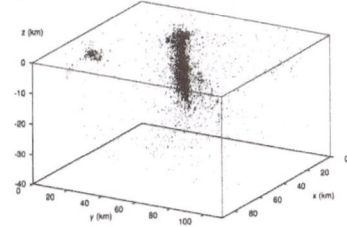

Fig. 8.3. Perspective views of the hypocentral aftershock distribution of the Hyogo-ken Nanbu earthquake at 30° (left) and 120° (right) rotation around the z-axis

8.1 Multifractal Properties

Spatial Distribution. A multifractal analysis of the epicentre distribution respectively its probability density distribution was carried out in analogy to previous similar calculations in this work. The D_q and $f(\alpha) - \alpha$ curves are shown in Fig. 8.5.

The error-bars indicate one standard deviation for the values of D_q which were directly determined. The values in between were obtained by cubic interpolation to enable the determination of a smooth $f(\alpha) - \alpha$ curve. One

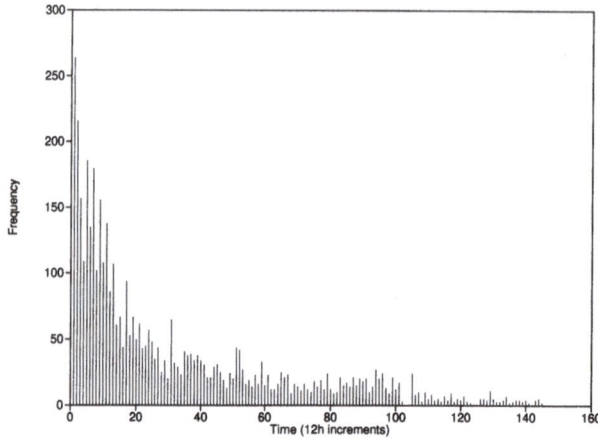

Fig. 8.4. Half-daily frequency of aftershocks of the Hyogo-ken Nanbu earthquake, showing the validity of the Omori law

notices a surprisingly small error, indicating a good fit, i.e. smooth linear scaling behaviour within the scaling limits. Indeed the narrowest scaling region (for $q = -8$) was still as large as 4.73 km to 53.57 km, which is sufficient to believe the fractal properties of the data. Δ was found to be 1.43±0.13 while Δ' was 1.00 ± 0.06.

The log − log plots exhibited a single linear region between the horizontal regimes, meaning that no multi-scaling behaviour was visible. The latter result might seem surprising because the linear elongated structure of the fault zone might well produce a crossover point (cf. [HIY92]). Here, however, the linear dimension of the fault is of the order of the whole area of analysis and thus no multiscaling becomes apparent.

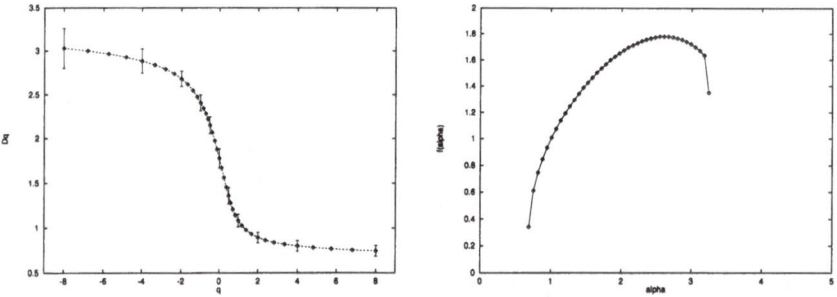

Fig. 8.5. Spectrum of generalised dimensions and multifractal $f(\alpha) - \alpha$ curve for the epicentre distribution of aftershocks of the Hyogo-ken Nanbu earthquake

The multifractal analysis of the hypocentre distribution, i.e. in three dimensional space, yielded practically the same results except for worse scaling behaviour, showing the need for more data points when increasing the "embedding dimension" and at the same time confirming that no information is lost when restricting the analysis of earthquake locations to a projection of the hypocentres to the two dimensional surface (cf. also Chapter 3).

Temporal Distribution. Due to the geometrically increasing number of necessary data points when increasing the "embedding dimension" (cf. Section 2.2.5), an analysis of space-time, i.e. with a fourth dimension time added to the three dimensional hypocentre distribution, is not feasible here (not to mention the complete "earthquake space", i.e. the five dimensional x, y, z, time and earthquake size space). Therefore a separate analysis of the temporal seismicity pattern has been carried out. Figure 8.6 shows a plot of inter-arrival times between successive aftershocks in the fashion of [GC95]. Two large values in the last third of the plot have been omitted for clarity. One notices a power-law like increase in the intervals, bringing to mind some kind of inverse Omori law.

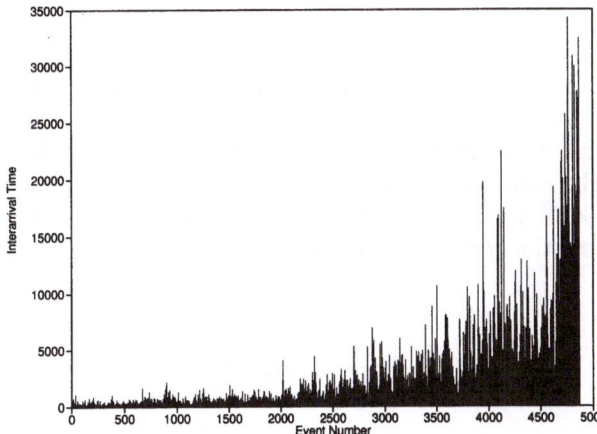

Fig. 8.6. Inter-arrival times for the Hyogo-ken Nanbu aftershock data

More instructive in the sense of fractal Cantor dusts (i.e. fractal distributions of points on a line) is Fig. 8.7, where the temporal distribution of aftershocks is shown in successive enlargements by a factor of 2. One expects the densest clustering to the left of every (sub)interval, decaying according to a power law to the right. The successive zooms, which always start with the first aftershock in the catalogue, show that this seems to be true up to a certain level when the scaling breaks down: the distribution becomes sparse near the starting time as well. The latter effect might be due to an inherent lower scaling limit or, more probably, due to the missing of events below a certain temporal resolution by the seismometer network respectively the evaluation.

The multifractal spectrum together with the D_q curve is shown in Fig. 8.8. Note that the $f(\alpha) - \alpha$ curve is for a one dimensional distribution, thus $\Delta\alpha$ cannot be directly compared with the one of two dimensional distributions. Both Δ (1.13± 0.14) and Δ' (0.43 ± 0.05), differ distinctly from the

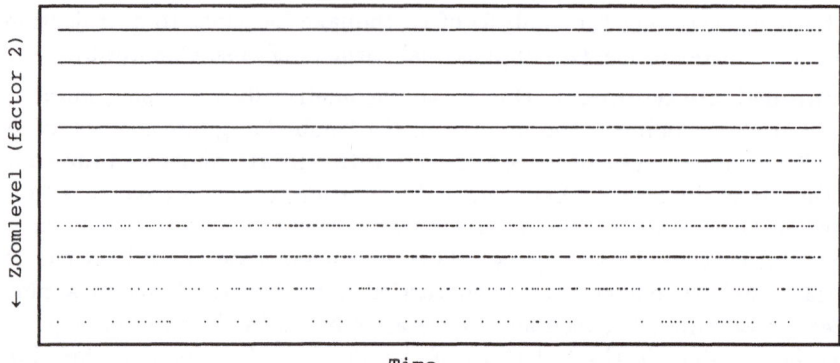

Fig. 8.7. One-dimensional "dusts" showing the temporal distribution of aftershocks at successive enlargements by a factor of 2. Time always starts with the first aftershock contained in the catalogue

monofractal value of zero. Thus one is lead to believe that the temporal aftershock distribution obeys multifractal laws as well. The numerical range of the input data (one dimensional occurrence times of the earthquakes as shown in Fig. 8.7) was 0 (time of first aftershock) to 6.310284×10^6 secs (time of last aftershock in catalogue). The scaling region in the $\log - \log$ plots ranged from about 640 secs to 3.449555×10^6 secs, i.e. the scaling breaks down below about 10 mins. The latter confirms the conclusions made from the examination of Fig. 8.7 above.

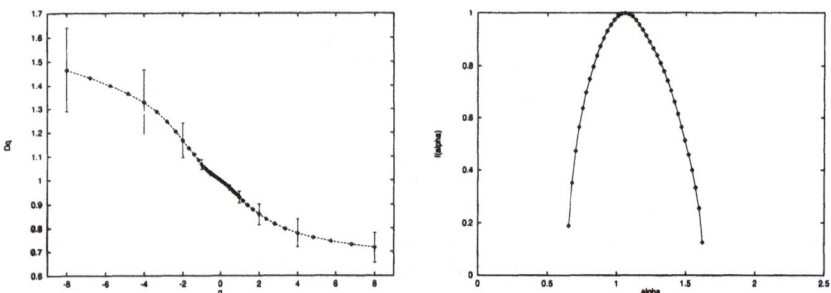

Fig. 8.8. Spectrum of generalised dimensions and multifractal $f(\alpha) - \alpha$ curve for the temporal distribution of aftershocks of the Hyogo-ken Nanbu earthquake

Size Distribution. Although a multifractal analysis of the earthquake size field (here: rupture areas) was again not found to be feasible, a one dimensional analysis of the size distribution was carried out in analogy to the previous section. Figure 8.9 gives the data used as input (actually the cumulative rupture area was used as input to have a one dimensional distribution on a line (a "dust")).

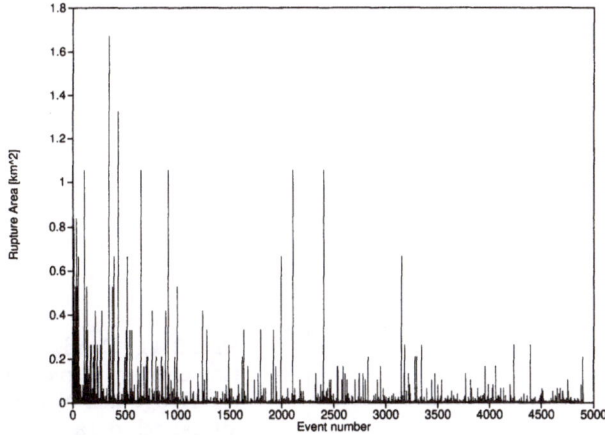

Fig. 8.9. Rupture area as a measure of earthquake size versus event number for the aftershocks of the Hyogo-ken Nanbu earthquake

The results of multifractal analysis are displayed in Fig. 8.10. For the cumulative rupture area measure, a very large scaling region of 1.61 to 59.75 km^2 was obtained (the final cumulative value in the input data was 74.99 km^2). The non-uniformity factor Δ resulted to be 1.18 ± 0.07, while Δ' was 0.48 ± 0.01, i.e. very similar to the result for the temporal distribution. It thus seems that the earthquake size distribution also constitutes a multifractal measure.

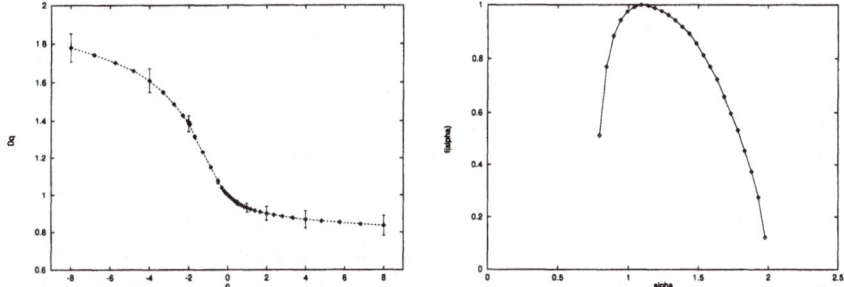

Fig. 8.10. Spectrum of generalised dimensions and multifractal $f(\alpha) - \alpha$ curve for the distribution of rupture area of aftershocks of the Hyogo-ken Nanbu earthquake

8.2 Configuration Entropy

A configuration entropy analysis was carried out for the aftershock data (epicentre distribution) to confirm and complement the previously obtained results. The result is given in Fig. 8.11.

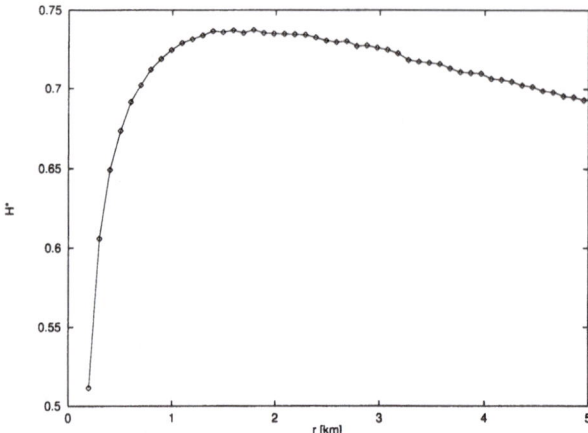

Fig. 8.11. Configuration entropy curve for the Hyogo-ken Nanbu aftershock data

One notices a single clear maximum at the optimal entropy length of $r^* \approx$ 1.9 km, confirming a multifractal structure of low or no lacunarity. The latter was to be expected due to the absence of (periodic) voids. When comparing this result for r^* with the one for the overall seismicity of the area analysed in Section 7.2.3, the value is found to be much higher. It is however, well within the limits of the moving results given in Fig. 7.17 as was to be expected because the data analysed there also contains aftershocks.

8.3 (An)Isotropic Fractal Properties

The results of a HOT analysis of the epicentre distribution and the distribution of cumulative rupture areas are shown in Figs. 8.12 and 8.13. Note that these fields are not probabilities because they are not normalised. Normalisation was not necessary because we are not interested in the scaling of moments. The field of cumulative rupture areas was generated from the given magnitude values as described in Section 7.2. Rupture area as a measure of earthquake size was again preferred to seismic energy or moment due to the reasons given in Sections 7.2 and 5.3.

Both analyses, i.e. of the distribution of epicentres ("location") as well as the one of rupture areas ("size"), show very similar results. Table 8.1 summarises the results. Most apparent is the orientation of the major axis of the ellipse in the rose plot ("angle"), i.e. the direction of highest fractal dimension: it is about 47° and thus agrees with the orientation of the fault line (about 48°, see Fig. 8.2) along which most of the aftershocks occurred. The fractal dimension in the direction of the fault line was about 1.85 for the epicentre distribution, while it was only about 1.04 in the perpendicular direction. The degree of anisotropy ("ratio") is very high at about 0.30 (recall that the greater the deviation from 1, the stronger the anisotropy) as was to be expected. For the rupture area field, it is still very high at 0.36 but the

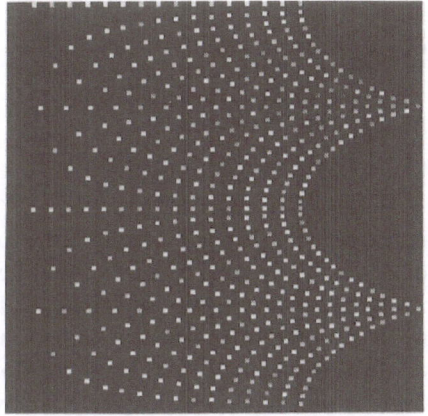 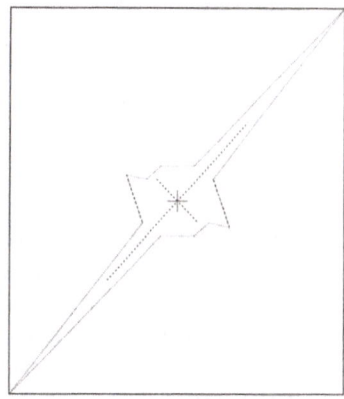

Fig. 8.12. HOT table and rose plot of H for the distribution of aftershock epicentres

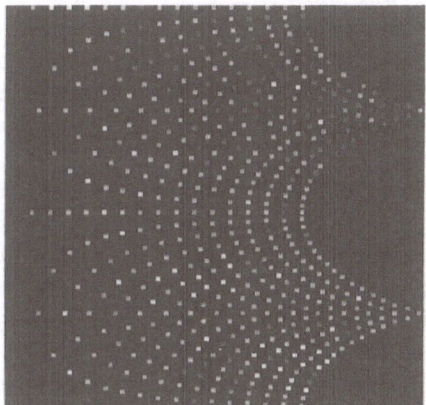 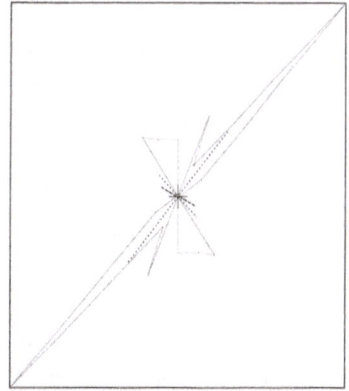

Fig. 8.13. HOT table and rose plot of H for the distribution of aftershock rupture areas

distribution of earthquake "size" is not as anisotrop as the distribution of epicentres. The aftershocks thus follow the linear elongated structure of the fault zone better in their location than in their size.

8.4 Overall and Aftershock Seismicity

As already mentioned, the catalogue used in Chapter 7 contains "normal" distributed seismicity as well as aftershock activity while the data analysed here consists of an aftershock sequence only. A comparison between the fractal

Table 8.1. Summary of the anisotropy analysis of the epicentre distribution and rupture size distribution

	Location	Size
Angle	46.61	47.61
Ratio	0.30	0.36
D	$\approx$ 1.04 - 1.85	$\approx$1.00 - 1.88

properties of the two sets, as far as meaningful with respect to the different numerical properties (number of data points etc.), will thus give an idea about the fractal differences between overall properties of seismicity and the special case of aftershocks. Such a comparison is of interest to further judge the discriminatory powers of fractal analysis. In the following, results of Section 7.2 and the current chapter are compared. It will be interesting to see how and to what extent the known differences in the degree of clustering and in the temporal distribution (Omori law) will be reflected.

First, the difference between the configuration entropy analyses is found to be significant in that $r^*_{overall} = 0.81$ km and $r^*_{aftershock} = 1.9$ km. The different values show that the two data sets display their maximum of disorder at quite different scales. When comparing Figs. 7.7 and 8.11 where the whole entropy curves are given, it can also be seen that the entropy value for aftershocks decays slower, here the entropy maximum at r^* is thus less pronounced for aftershocks.

Next, there may be noticed a great difference between the fractal anisotropies of spatial density as well as earthquake size distribution as determined by HOT analysis: despite the "spikyness" of the rose plots shown in e.g. Figs. 7.5 and 8.12 for overall and aftershock seismicity, respectively, the aftershock fields are correctly reported to possess much stronger anisotropy. The actual values are reported in Table 8.2.

Most rewarding is maybe the comparison of the multifractal properties of both the temporal and epicentre distributions of the two seismicity states. Considering the one-dimensional temporal distributions first (cf. Figs. 7.4 and 8.8), it can be seen that the multifractal spectrum of aftershocks is not only much less homogeneous but that the number of sparsely populated intervals in time has decreased dramatically—$f(\alpha_{max})$ decreased to less than 0.2 as compared to greater than 0.8 for the overall seismicity. A less pronounced change may also be observed in $f(\alpha_{min})$ which increased from about 0.03 to almost 0.2. The number of most densely clustered intervals has thus decreased a little. The increase in heterogeneity of the fractal distribution of earthquakes in time is mainly due to a simultaneous increase in α_{max}. Thus not only the number of sparse time intervals is reduced (which was to be expected) but the degree of sparsity of the latter few intervals increases. The latter is not contradictory as there are extremely few sparse intervals as compared to overall seismicity. It is also noteworthy that α_0 and $f(\alpha_0)$ are not affected.

Significant changes also occur in the multifractal spectra of the epicentre distributions, i.e. the spatial density (see Figs. 7.3 and 8.5). From overall seismicity to aftershock seismicity, the greatest change also occurs in $f(\alpha_{max})$: it increases from about 0.5 to about 1.3, indicative of the increase in extremely sparse vicinities, i.e. areas of extremely low seismic activity. These sparse areas within the spatial aftershock distribution are much more densely populated than those of the overall seismicity as witnessed by the decrease in α_{max} from about 4.8 to about 3.3. The decrease in heterogeneity is mainly due to the change in α_{max} as α_{min} stays almost constant. A difference may also be noticed in the values of α_0, which decreases. Most clusters therefore become more densely populated during aftershock activity. $f(\alpha_0)$ decreases slightly from ≈ 1.95 to ≈ 1.8. As $f(\alpha_0) = D_0$, this denotes a decrease in the density with which the epicentres fill the two-dimensional plane. The latter was also to be expected (cf. Figs. 7.1 and 8.2).

The transition from overall to aftershock seismicity may thus be characterised by an increase in spatial homogeneity with a simultaneous decrease in temporal heterogeneity. Table 8.2 summarises some of the above mentioned differences.

Table 8.2. Some significant differences between fractal parameters of overall and aftershock seismicity

	Entropy spatial r^* [km]	Anisotropy spatial factor	size
Overall	0.81	0.60	0.50
Aftershock	1.90	0.30	0.36

		Multifractal spatial			temporal		
	Δ	α_{max}	$f(\alpha_{max})$	Δ	α_{max}	$f(\alpha_{max})$
Overall	2.09	4.8	0.5	0.53	1.1	0.87
Aftershock	1.43	3.3	1.4	1.13	1.6	0.12

To conclude the comparison of overall seismicity and aftershock sequence, it may be said that fractal parameters are useful to distinguish them. The latter again confirms the ability of fractal analysis to detect anomalous and therefore possibly precursory seismicity.

Part II

Earthquakes and Chaos

9. Chaos

"Chaos" here always means low-dimensional deterministic chaos. Deterministic chaos names the irregular behaviour of a nonlinear system whose temporal evolution is completely determined (by mathematical equations). Almost identical initial conditions (or small perturbations during iteration) lead to exponentially diverging solutions. The latter is the reason for very limited predictability of chaotic systems—due to measurement error, even at the Heisenberg uncertainty level, it is impossible to predict the trajectory of even a mathematically completely known chaotic system for all times. Using classical linear methods of time series analysis, most chaotic systems can in fact not be distinguished from "noise", i.e. random signals without underlying determinism.

"Chaos" or nonlinear science has become a vast field of research, greatly surpassing fractal geometry. Most research in nonlinear science, however, deals with simulation and subsequent analysis of nonlinear models represented by differential or difference equations, maps or cellular automata. Other fields comprise, e.g., the application of chaos in engineering (e.g. [Moo92]) or the control of chaos (e.g. [AFH94]). The number of publications in such a variety of fields as biology, economics, sociology, medicine, geosciences, history, mathematics and naturally all fields of physics dealing with chaos has reached more than 10 000 since 1980.

In geophysics, several papers including studies of the magnetosphere (e.g. [Rob91] which also includes a good introduction to delay-time embedding, [SVP93]) or, closely related, mantle convection ([Tur92] and references therein, [PJ94]) and the earth's magnetic field may be found. Works directly related to earthquakes include the study of cellular automata (e.g. [OFC92, Nak90, CJV92, MT91]) which are usually discrete slider block (stick-slip) models, the analysis of analytical sliding block models ([Tur92]) and general discussions of the question whether the earthquake process should be considered chaotic or not (e.g. [HT90], [Sch89] where the conference "Earthquakes: chaotic or deterministic?[1]" is discussed). A review article by Carlson *et al.* (1994) comprehensively discusses the ongoing research into the Burridge-Knopoff model of earthquake faults in the context of current seis-

---

[1] Asilomar, California, 12-15 February 1989; the title of the conference is unfortunate as deterministic chaos is usually meant when the term "chaos" is used

mology, including generalisations towards more realistic higher-dimensional models and implications of their chaotic behaviour for prediction.

Cellular automata replace systems of coupled nonlinear differential equations when the latter would be too complex to solve (analytically and/or numerically) and have led to the conclusion of chaoticity of earthquakes in some cases. For an overview and references see e.g. Main (1996) and Rundle *et al.* (1996), the latter especially for latest developments such as the Traveling Density Wave model for earthquakes. An important paper by Grassberger [Gra94], however, points out that the observed chaotic behaviour of discrete slider block cellular automata might well be due to numerical limitations or the wrong choice of boundary conditions. Also the analytical modelling of a slider block system as carried out by Turcotte *et al.* (1993) leaves the question of adequacy of the low-dimensional analogon of nature (only a limited number of blocks, as low as two, is used in the simulations). Thus such models will never be able to prove or defy chaoticity of earthquakes. Indeed, also the limitation of (numerical) nonlinear analysis to low dimensionality might appear as a severe limitation in the case of the earthquake process which must be regarded as extremely complex due to its analogy to turbulence alone. In many cases in nature, however, the asymptotic behaviour of infinite-dimensional systems is effectively low dimensional (e.g. [Tso92] where it is demonstrated that systems of partial differential equations which possess an infinite-dimensional phase space can settle on low-dimensional attractors; the latter is especially true where dissipation plays an important role—as in the case of earthquakes). The phase space of a dynamical system is a mathematical space with orthogonal coordinate directions, one direction for each of the variables necessary to specify the instantaneous state (phase) of the system. A trajectory in phase space is produced by plotting the systems evolution with time. A sufficiently long trajectory generates the phase space diagram which in the case of chaos forms a strange attractor.

So far, observation of supposedly chaotic earthquake behaviour in the real world may only be regarded as evidence (see e.g. Huang and Turcotte (1990) for an example where actually observed chaoticity is claimed). Also the observation of fractal statistics of earthquakes which, beyond any doubt, can be seen as such evidence as will be detailed below (cf. also [Tur92]). Another interesting point is the question of chaotic mantle convection (as expressed by, e.g., magnetic field reversals) as a driving force in plate tectonics (see e.g. [Tur92]).

As pointed out by Meissner (1994), two sources of nonlinearity in the initiation and continuation of rupture processes are the build-up of critical stresses and their propagation along inhomogeneous rupture zones, including the non-seismogenic lower crust. According to this author, the most important sources for a nonlinear build-up of critical stresses seem to be dilatancy processes, transfer of stress from nearby foreshocks, and possibly interactions with the lower crust. Inhomogeneities of dynamic friction, asperities, fault gouge and geometry all influence propagation of the rupture process. An important conclusion by Meissner is that the deterministic, i.e. non-probabilistic, prediction of earthquakes would require a dense network as well as short- and long-term monitoring of any deformation in the neighbourhood of the suspected rupture area. A major argument towards the nonlinear build-up of critical stresses is the fact that large earthquakes show no correlation with the earth's tides ([Kno64, RSSS93]). Also the fractality of faults (e.g. [Hir89b]) at all scales, including the fractal distribution of gouge material itself ([Ble91]) introduce nonlinearity into the rupture process itself—another fact not reflected in slider block models. In conclusion, several arguments for nonlinearity or against linearity in the stress build-up and in the rupture process itself may be found.

Nonlinearity is necessary but not sufficient for chaotic behaviour, however (e.g. [Ott93]). Therefore, only the proof of chaotic determinism in earthquake-related real world time series could really answer the question of chaoticity in earthquakes. While time series can be derived from earthquake catalogues (e.g. earthquake intervals, see below), interest here is also on two directly observed geophysical "earthquake time series": radon emission and strain. The latter two signals represent a choice of two typical parameters commonly thought to be signals related to crustal processes leading to earthquakes. Because of that they are frequently monitored in experiments throughout the world to detect earthquake precursors. So far, success has been very limited (e.g. [Gel97]). The persisting problem with these and other geophysical field observations is the unambiguous detection of anomalies which could be safely regarded as precursors. Despite the inherent nonlinearity of the signals, detection of precursors is attempted by linear methods up to now. Detection of precursors by nonlinear means and a discussion of implications of the findings for chaoticity of the earthquake process are attempted in this part. Concepts and terms from nonlinear time series analysis will be introduced when needed.

9.1 Nonlinear Time Series Analysis

Nonlinear time series analysis is very much the object of current research and as such no general rules or procedures exist for the details. For the overall approach, however, the current consensus is roughly outlined in Fig. 9.1. More detailed explanation and description of the individual steps follows below as they are applied. It should be noted that further further methods not shown in Fig. 9.1 exist. Those methods, such as, e.g., nonlinear prediction in phase space, either complement or confirm the steps shown. See also Section 9.2.1.

A central idea in reconstructing nonlinear dynamics from a scalar time series is the one of delay time embedding (e.g. [Ott93]), to be briefly introduced here. Let the d-dimensional vector $\boldsymbol{x}(t)$ be the state vector, i.e. the phase a dynamic system with d degrees of freedom is in at time t. In geophysical field observations, one doesn't know d, and even if one did, one could probably

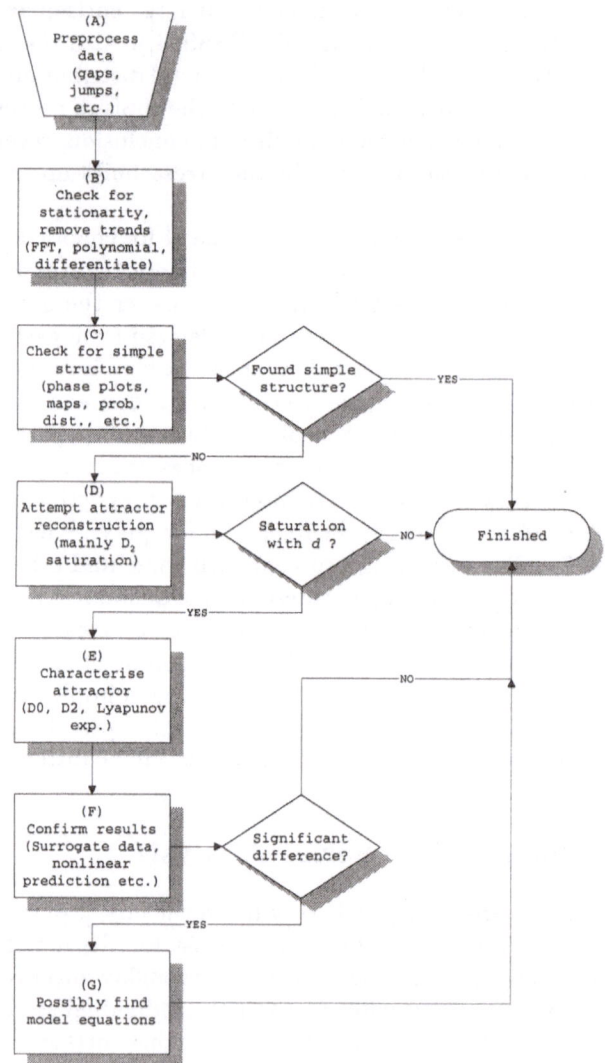

Fig. 9.1. Flow diagram of nonlinear time series analysis strategy

not measure all of the components of x simultaneously. Instead, we measure one scalar function of the state vector,

$$g(t) = G(x(t)),$$

a continuous time series such as strain (nobody would seriously assume that strain is a scalar, or, more important, that the dynamics which underlie the strain signal are one-dimensional).

Thus, one must somehow try to reconstruct the full dynamics from $g(t)$. In phase space, this means a reconstruction of the attractor, the geometric object the dynamics are confined to (the attractor is made up of the trajectories of the system). The attractor may be a point, a cycle (for a periodic system) or a strange, i.e. fractal, object (chaos). That it is indeed possible to reconstruct the attractor from one scalar time series is surprising but has been shown by Takens (1980). Takens showed that the attractor can be obtained as follows. Define the delay coordinate vector $y = (y^{(1)}, y^{(2)}, \cdots, y^{(d)})$ by

$$
\begin{aligned}
y^{(1)}(t) &= g(t) \\
y^{(2)}(t) &= g(t - \tau) \\
y^{(3)}(t) &= g(t - 2\tau) \\
&\vdots \\
y^{(d)}(t) &= g(t - (d-1)\tau)
\end{aligned}
$$

where τ is an appropriate delay time (see below). $y(t)$ may then be regarded as a function of $x(t)$, $y = H(x)$. If the number of delays d (or: the embedding dimension) is sufficiently large, y-space (the reconstructed attractor) is qualitatively equivalent to the original phase space. A first idea about possible dynamics behind $g(t)$ may be gained from simply examining the continuous time trajectory in y by plotting e.g. $g(t)$ versus $g(t-1)$. Such plots will be shown later on.

Figure 9.2 explicitly demonstrates the above mentioned finding by Takens (1980), and thus the method of delay-time embedding, to discover the dimensionality of a possible attractor from a single scalar time series. The central realisation is that it is often possible to represent the state of a d-dimensional system by independent variables (coordinates) other than the "physical" state variables (which are a priori unknown here) as a function of time. If one assumes a deterministic system, i.e. different initial conditions lead to unique solutions, any trajectory of the d-dimensional system may also be specified by the coordinates of only one variable at d different times. In Fig. 9.2, the two-dimensional case is shown. The physical coordinates of the system are x and y, e.g. position and velocity of a particle moving in one dimension (a particle moving in three dimensions requires a six-dimensional phase space, three coordinates for position and three for velocity). The three curves α, β and γ represent three different trajectories in the phase plane; The system's state at time t is indicated by an asterisk, the state at time $t + \tau$

is marked by "o". It becomes obvious that trajectory β may be equally selected (described) by specifying that it passes through $x_d(t)$ and $x_d(t+\tau)$ (the delay coordinates) as it passes through the point $x_d(t)$ and $y_d(t)$; trajectories α and γ have evolved somewhere else at time $t + \tau$.

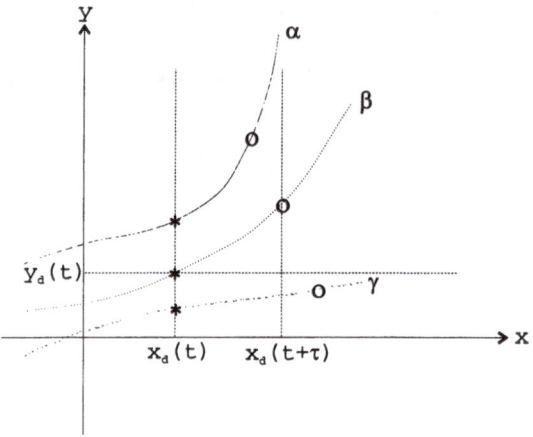

Fig. 9.2. The Takens theorem of delay-time embedding: Trajectory β may be equally well specified by saying that it passes through the state $x_d(t), y_d(t)$ or by saying that it passes through $x_d(t)$ and $x_d(t + \tau)$

Note that adding a third delay coordinate $x_d(t + 2\tau)$ would still specify the trajectory but that this would be redundant. Thus, in this case of two physical state variables, only two delay coordinates are required. As this is the case for any d-dimensional system, finding the number of necessary delay coordinates tells the number of physical state variables needed to model the system under analysis. Redundancy of adding additional delay coordinates can be detected by looking for saturation in a plot of e.g. the correlation dimension of the phase diagram (attractor) versus embedding dimension (see below). The value of the embedding dimension (number of orthogonal delay coordinate directions) at which D_2 saturates is thus a lower estimate for the number of variables (and thus equations) needed to model the system. The upper limit of variables to describe the dynamics of a system evolving on a D-dimensional attractor is $2D' + 1$ where D' is the next highest integer of D when D is fractional ([Tak80]). In practice, $D' + 1$ variables are often sufficient to spread the attractor out in phase space and make it single valued (as is required for a deterministic system).

The fractal dimension of the attractor is a lower bound for the number of variables (number of e.g. coupled nonlinear differential equations) needed to model the process. It is thus of great interest to determine the fractal dimension of the assumed attractor of the dynamics of unknown natural systems to get an estimate for their dimensionality. The latter is the first step in modelling the underlying dynamics of all observed geophysical time series. It also explains, why one-dimensional representations (i.e. plots of some quantity versus time) are inadequate to represent the dynamical behaviour of the process observed and motivates the idea that possible precursors will not be visible in such a representation. Thus, even if observed geophysical signals do not stem from chaotic systems, they are certainly not one-dimensional and must be examined in higher dimensional space to unveil significant changes in behaviour or significant anomalies.

9.2 Analysis of Known Dynamics

Before attempting to analyse geophysical field data, three examples of different classes of dynamics will be examined using the methods indicated above. This will make it possible to introduce the methods in more detail, show their advantages and shortcomings and contribute to the understanding of the aim of nonlinear time series analysis as conducted here. In particular, a non-trivial periodic system, an infinite-dimensional signal (Gaussian noise) and finally a classical example of low-dimensional chaos (the Lorenz system) will be examined. Throughout this section, the idea is to show to what conclusions the analyses would lead if the dynamics of the signals were unknown. The latter reflects the situation when analysing geophysical field data.

9.2.1 Quasi-periodic Dynamics

Figure 9.3 shows 2000 points of a seemingly random signal which in fact is simply the superposition of two incommensurate (non-integrally related) harmonic functions: $y(t) = \sin(t/2) + \cos(gt/2)$ where $g = (\sqrt{5} - 1)/2$ is the inverse of the golden mean and $0 \leq t \leq 1999$ (upper curve, offset by $+4.0$). This is a standard example given in many textbooks on chaos to show the effect of unveiling structure in phase space as opposed to the apparent randomness in the classical one-dimensional "time series plot". The lower curve is a randomised (shuffled) surrogate data set derived from the original data by simply shuffling the values randomly; thus the probability distribution is preserved, but the power spectrum and autocorrelation function are altered. The surrogate data set will be required for confirmation of results further below. For now it can serve for judgement by optical inspection whether the original data displays more structure than the randomised signal or not.

Following the strategy outlined in Fig. 9.1 (step B), Fig. 9.4 gives the power spectrum of the original data (as also in the following, the logarithm of power normalised to unity is plotted versus frequency in units of the Nyquist frequency, i.e. $1/2\Delta t$, see Section 9.3.3 for numerical details of how the spectrum is obtained).

The two frequency constituents are immediately recognised, the power spectrum is not broad ("noisy") and shows no power-law behaviour. The latter two properties would have made the data a candidate for chaos (see below). The trailing zig-zag curve towards higher frequencies is a spurious

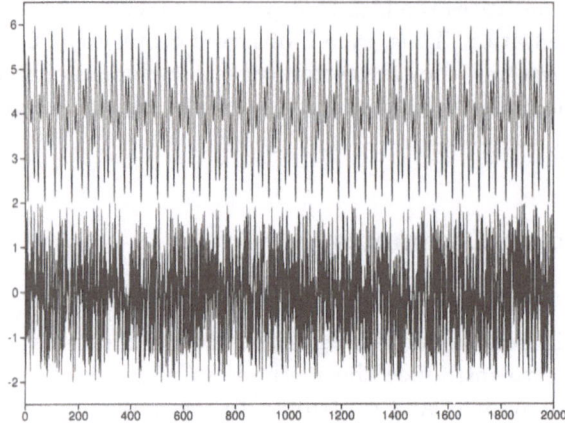

Fig. 9.3. A seemingly random signal of two incommensurate frequencies (above) and its shuffled version (below)

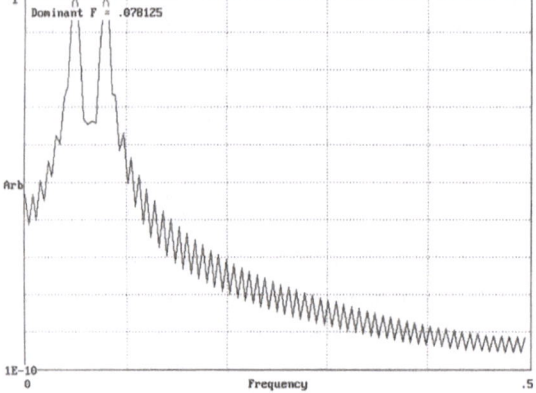

Fig. 9.4. Power spectrum of quasi-periodic data

response due to the high number of frequency intervals (128) used for analysis to achieve high resolution.

The above finding already corresponds to step C in that at this early stage simple structure is revealed (by the linear method of Fourier analysis). One might conclude to be dealing with a linear quasi-periodic system (the two narrow peaks are not obviously in an integer ratio) which may be modelled sufficiently by the superposition of harmonics. No further nonlinear analysis is required, as the possibility of chaos has been excluded already. The assumption of quasi-periodicity may be easily confirmed by looking at the autocorrelation function which is given in Fig. 9.5: for (quasi-)periodic functions, the autocorrelation function obviously does not stay at zero and decays only very slowly, if at all, with τ.

To be able to demonstrate some of the further methods and to be able to compare with later results, Figs. 9.6 and 9.7 show a stereographic phase space plot of $x(t)$ versus $x(t - 1)$ (i.e. a plot of two delay time coordinates with delay time $\tau = 1$) and a return plot. The latter plots (also called return

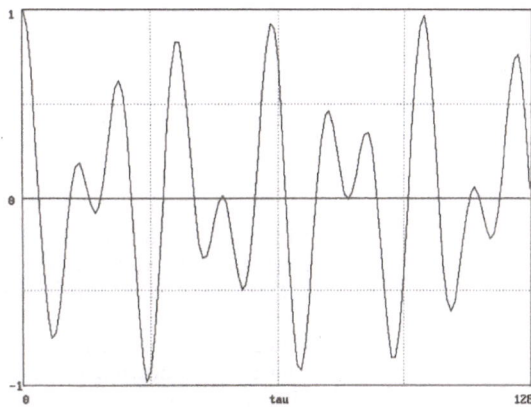

Fig. 9.5. Autocorrelation function of quasi-periodic data

maps) are often superior to simple phase space plots in distinguishing noise and chaos as they are a kind of cross-section of the phase plane and thus reduce the the dimension by one. The name return map stems from the fact that instead of the values itself, recurrences of certain conditions are plotted versus each other. In Fig. 9.7, $x(t)$ is plotted at positions where $x'(t) = f$ (y-axis) versus the previous time where the condition was fulfilled (x-axis). Depending on f, this procedure is more or less revealing. Here, $f = 0.50$.

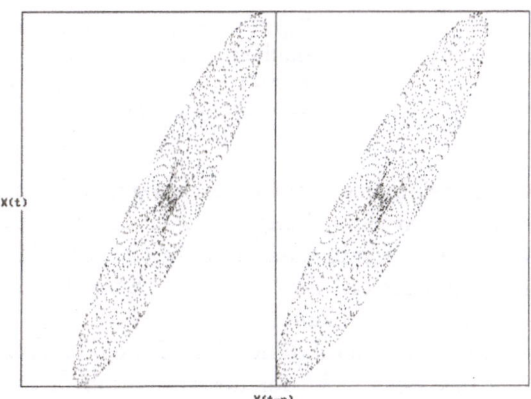

Fig. 9.6. Stereographic delay time plot of quasi-periodic data revealing a torus

The phase space plot unveils that the "dynamics" lie on the by now classical two-torus; note the difference in information unveiled by one-dimensional scalar plots and phase space plots! The return map may easily be recognised to be a cross-section of the torus. Both plots additionally confirm the simple structure of the data concerned. Fig. 9.8 demonstrates the effect of randomisation in phase space by embedding the shuffled data shown in the lower part of Fig. 9.3: the structure has been completely lost.

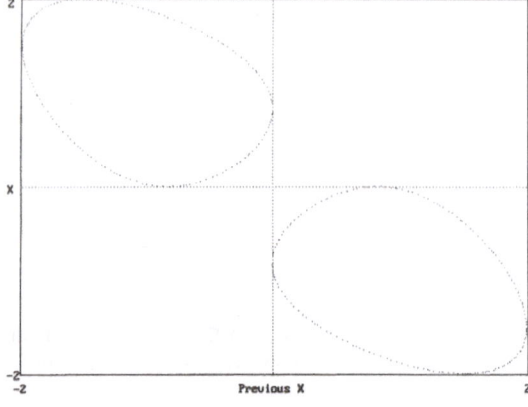

Fig. 9.7. Return map of quasi-periodic data in which $x(t)$ is plotted at positions where $x'(t) = 0.5$ (y-axis) versus the previous time where the condition was fulfilled (x-axis)

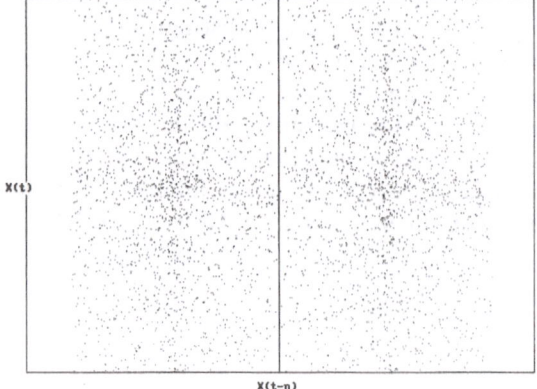

Fig. 9.8. Phase space plot of shuffled quasi-periodic data: The structure seen in Fig. 9.6 has been destroyed

Final confirmation of the low-dimensionality of the data comes from steps D and E in Fig. 9.1, in this case the delay embedding of the scalar data in successively higher dimensions and the determination of D_2 of the resulting phase portraits for each embedding dimension. Fig. 9.9 gives the resulting function $D_2(d)$ where d is the embedding dimension.

One sees a clear saturation at $d \approx 3$, the fractal dimension of the attractor is about 2.5. The error bar for D_2 (see Sect. 9.3.3 for determination of the error in D_2) includes a range of about 2 to 3, indicating that the dynamics can probably be modelled by as few as 3 to 4 equations. Relying on the latter step alone would thus overestimate the complexity of the data (which might be due to the extremely small number of data points used here). The outcome of step F is shown in Fig. 9.10, where $D_2(d)$ for the surrogate data given in Figs. 9.3 and 9.8 is shown.

As was to be expected, no saturation with d is present any more, the structure in the original data has been destroyed. Note that the error bars also increase dramatically, thus indicating poor scaling properties of the randomised data. Both observations confirm that the original data possessed real

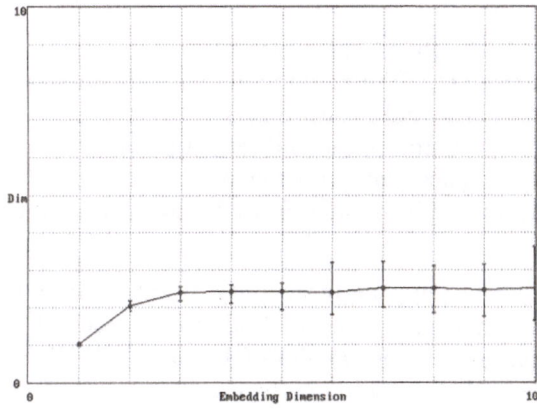

Fig. 9.9. $D_2(d)$ for embedding dimensions of 1 to 10 for quasi-periodic data

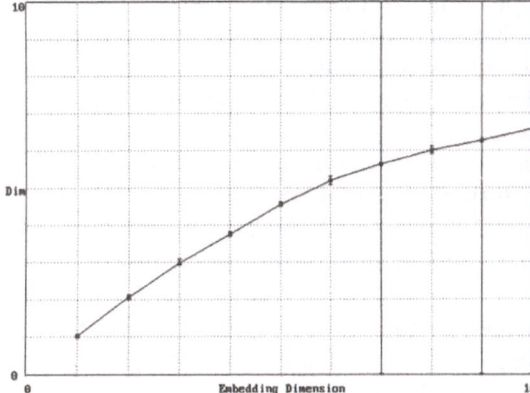

Fig. 9.10. $D_2(d)$ for embedding dimensions of 1 to 10 for shuffled quasi-periodic data for confirmation of structure in the original data

deterministic structure, i.e. that the saturation of $D_2(d)$ was not spurious. Spurious saturation happens for example in the case of coloured noise (see also below) when saturation persists even after randomisation.

Summarising the above results, the analyses would lead to the following conclusions:

- The data is low-dimensional ($D_2(d)$, phase and return plots)
- The data does not represent noise ($D_2(d)$ in conjunction with randomisation, no broad power spectrum, phase and return plots)
- The data is not chaotic (power spectrum, phase and return plots)
- The data is quasi-periodic (phase space plot is a torus, no circle)
- The data represents two superimposed incommensurate harmonics (power spectrum)

Further support could be obtained from the probability distribution and related tests (see below). Final and most powerful confirmation could be achieved by actually modelling and predicting the data (this is usually done by devising the model, deleting several of the final points, making a prediction for these deleted points from the model and then comparing prediction and actual values). Here the model could be obtained from a more detailed Fourier analysis in which also the phases were kept. The latter is possible only because the signal does not stem from a nonlinear model. In the case of nonlinearity, nonlinear prediction has to be employed ("data-implicit modelling", cf. e.g. [KG95] and below).

9.2.2 Infinite-dimensional Dynamics (Noise)

Fig. 9.11 shows 2000 random points with a Gaussian (normal) distribution with zero mean and a standard deviation of unity (cf. [P+92]). Depending on the quality of the random number generator[2], one would thus expect the total absence of determinism, i.e. an infinite-dimensional phase space. The data set scrutinised in this section thus represents the opposite extreme to the quasi-periodic example above.

Figs. 9.12 and 9.13 give a phase space plot and return map, respectively, in analogy to the section above. Both plots reveal no structure in that they fill the plane rather homogeneously. One may conclude that the data does not stem from a small number of discrete modes, no low-dimensional determinism is discernible.

The latter conclusion is confirmed by the probability distribution shown in Fig. 9.14 which is a simple Maxwellian curve (the normalised probability of occurrence of values is shown, see also below). Chaotic data would usually produce a fractal distribution, periodic data would possess sharp peaks. Another probability-related analysis is the IFS-clumpiness test (cf. e.g. [GW88]

---

[2] The methods detailed in this chapter may in fact be utilised to detect shortcomings in algorithms for random numbers

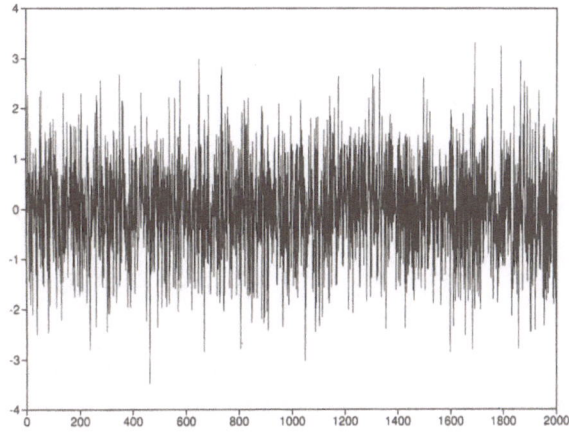

Fig. 9.11. A data set of 2000 Gaussian random points

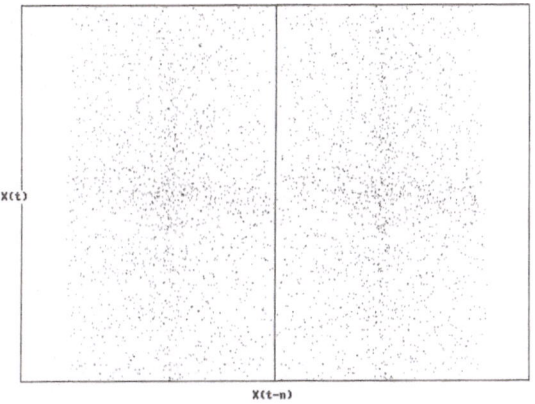

Fig. 9.12. Stereographic phase space plot of Gaussian noise

for iterative function systems, [SR95] and references therein for their application in the current context). The latter test produces uniformly distributed points in the plane for random data while chaotic data and coloured noise produce localised clumps. The result of such a test applied to the noise data here is given in Fig. 9.15. No structure is revealed either and the randomness of the data is confirmed.

The autocorrelation function in Fig. 9.16 drops to zero abruptly and remains there, also indicative of the absence of determinism (no correlation between successive values). The power spectrum (Fig. 9.17) is broad, no dominant periodic processes can be detected. In addition, it may be observed that the spectrum is flat, most probably excluding the possibility of chaos. In the case of flat, broad power spectra it is of interest to integrate the data to possibly generate a random walk (Brownian process, cf. Chapter 5) and determine the Hurst exponent of the process. Fig. 9.18 shows the resulting Hurst plot. The obtained value of $H = 0.498$ is characteristic of Brownian motion where

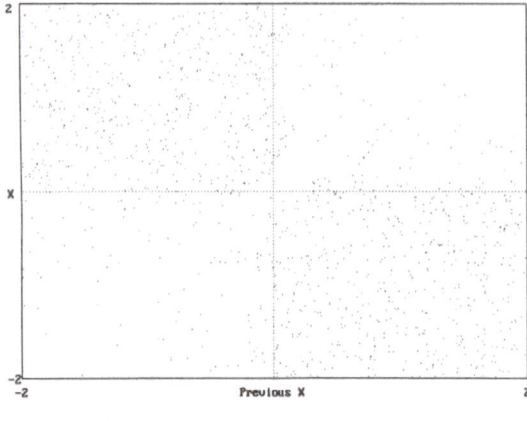

Fig. 9.13. Return map of Gaussian noise

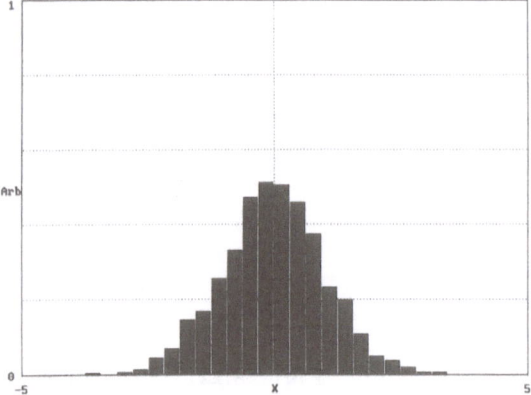

Fig. 9.14. Probability histogram of Gaussian noise

no persistent trend exists in the data. Thus the data is confirmed to be white noise.

Finally, the $D_2(d)$ calculation, given in Fig. 9.19, shows no tendency to saturate and the scaling properties are poor. Due to the definition of random data (infinitely many degrees of freedom), $D_2(d)$ should be of the order of d for all values of d but here only 2000 data points are analysed which leads to an underestimation of D_2. Taking the error bars into account, however, the latter requirement might well be fulfilled. If the data is random, further randomisation by shuffling should not alter the structure in phase space. This is confirmed in Fig. 9.20 which shows $D_2(d)$ for the shuffled random data. Indeed, the latter result is barely distinguishable from Fig. 9.19, thus confirming the randomness of the system. No further analysis nor deterministic modelling of such a signal is possible (with the methods used here).

To sum up the findings of this section one may say that

– The data possesses no simple structure (all tests)
– The data does not come from a small number of modes (power spectrum)

Fig. 9.15. IFS-clumpiness test of Gaussian noise

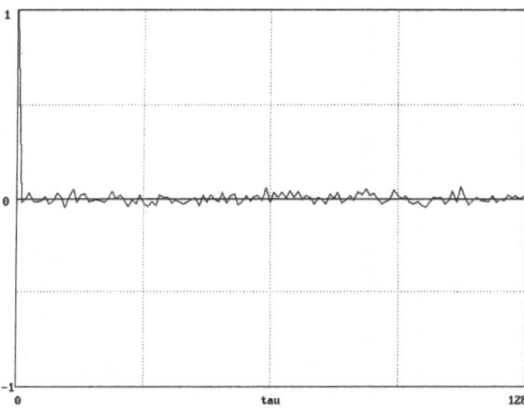

Fig. 9.16. Autocorrelation function for Gaussian noise

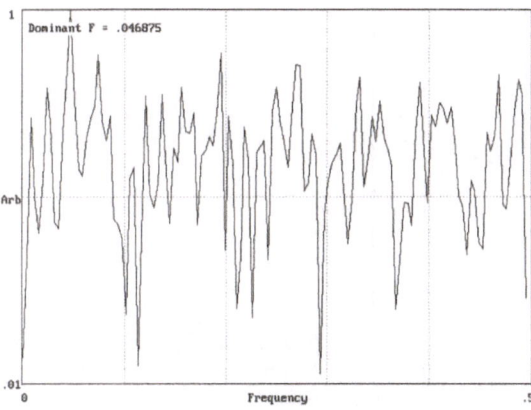

Fig. 9.17. Power spectrum for Gaussian noise

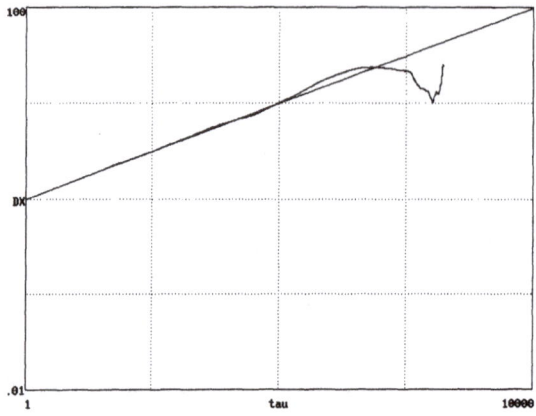

Fig. 9.18. Hurst plot for integrated Gaussian noise yielding $\dot{H} \approx 0.5$

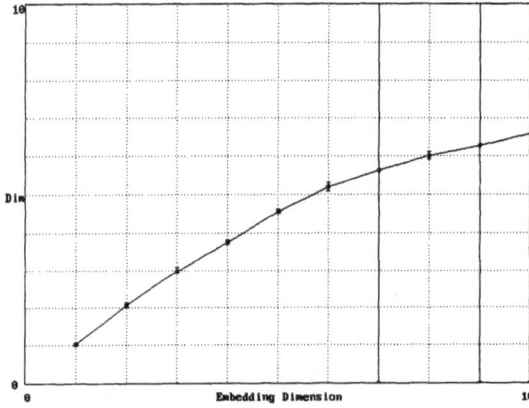

Fig. 9.19. $D_2(d)$ for $d = 1$ to 10 for Gaussian noise

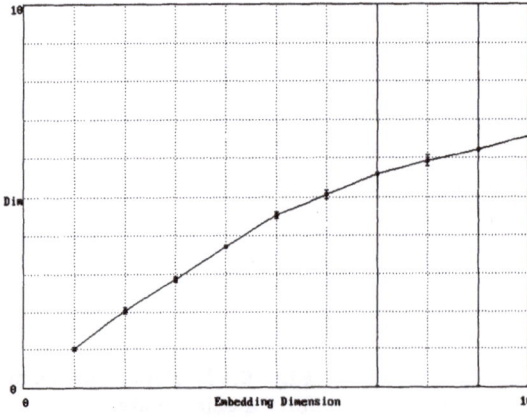

Fig. 9.20. $D_2(d)$ for $d = 1$ to 10 for "randomised Gaussian noise"

- The probability distribution looks Gaussian
- The integral of the data is a Brownian motion, i.e. the data represents white noise
- The autocorrelation function shows no correlation
- D_2 does not saturate

Hence, one must assume that the data possesses no low-dimensional determinism at all, whether chaotic or not. Although the system might be deterministic in a high dimension, it essentially represents noise to the (numerical) methods of analysis and modelling available currently. The Hurst exponent of about 0.5 furthermore strongly suggests that there is no determinism in higher dimension either. There is no way to model and thus predict such a system besides probabilistic methods.

9.2.3 Low-dimensional Chaotic Dynamics

The last data set considered in this section was generated by integrating the now classical Lorenz system of three differential equations

$$
\begin{aligned}
dX/dt &= 10(Y - X) \\
dY/dt &= 28X - Y - XZ \\
dZ/dt &= XY - 8Z/3
\end{aligned}
$$

2000 data points were obtained by using a sample interval of $\Delta t = 0.05$. The Lorenz attractor is described in almost any book on chaos and is a simplified model of a dissipative chaotic flow which was used by Lorenz (1963) to model atmospheric dynamics. Maybe the most comprehensive (mathematical) treatment of the Lorenz attractor may be found in Sparrow (1982). Turcotte (1992) gives a very detailed treatment including derivation of the equations in the context of mantle convection.

The signal is shown in Fig. 9.21 (upper curve) together with its shuffled version (lower curve) which will be needed for verification of results later on. Although a very different appearance of the original data and its shuffled version may be noticed, it is not obvious how the apparent structure in the original data should be described—the data seems neither completely random, nor periodic. Linear methods can indeed usually not distinguish between chaotic and random data.

Following the procedure of the sections above and the strategy outlined in Fig. 9.1, Figs. 9.22 and 9.23 give a stereographic phase space plot and a return map for the data. Both plots immediately reveal a well defined structure in the data. As mentioned earlier, animated sequences can not be reproduced here, but it should be pointed out that watching the evolution of the trajectory in phase space is an important additional means to distinguish deterministic chaos from e.g. coloured noise; the ways in which the attractors

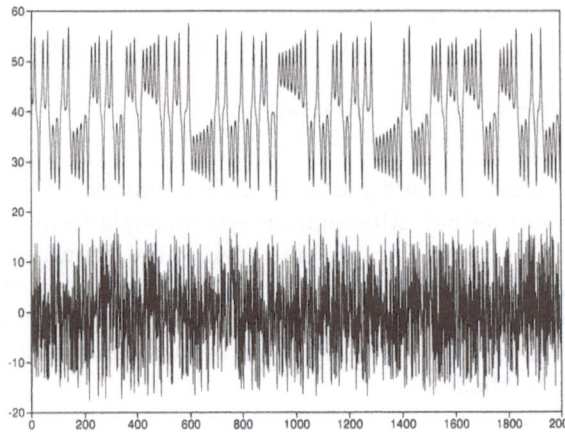

Fig. 9.21. A chaotic signal from the Lorenz system (above) and its shuffled version (below)

are built up differ significantly (cf. also [Rob91]). The return map is almost one-dimensional, implying that the attractor should have a fractal dimension of about 2.

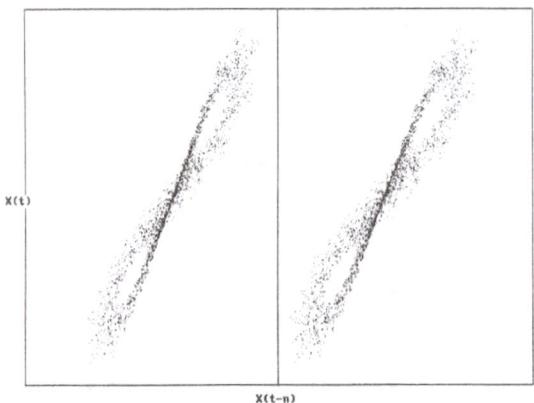

Fig. 9.22. Stereographic phase space plot of chaotic Lorenz data

The probability distribution in Fig. 9.24 is no simple curve but also possesses no sharp peaks as would be expected from a periodic system. Instead, the distribution is rather irregular, possibly fractal, indicative of a chaotic system. The IFS-clumpiness test shown in Fig. 9.25 clearly rules out random data in that the attractor of the IFS is extremely localised.

The power spectrum, displayed in Fig. 9.26, is broad and shows a very clear 1/f behaviour, i.e. a power-law relationship exists. Together with the absence of clear dominant frequencies, the power spectrum thus rules out the possibility of a quasi-periodic solution as dealt with in Section 9.2.1. A quasi-periodic solution would have been possible due to the assumed attractor dimension close to 2.

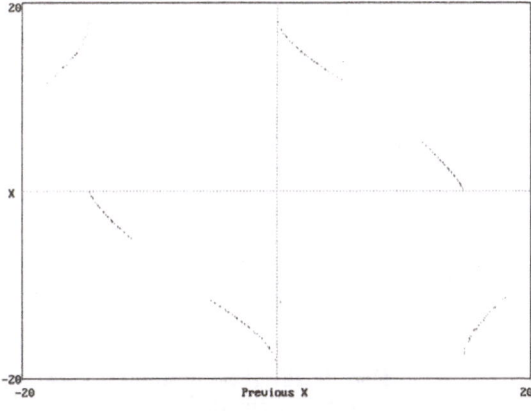

Fig. 9.23. Return map for chaotic Lorenz data

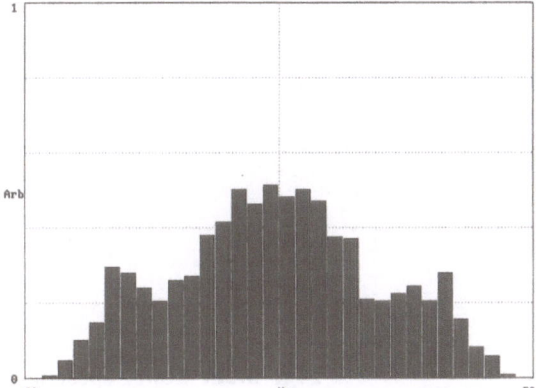

Fig. 9.24. Probability distribution of chaotic Lorenz data

Finally again, $D_2(d)$ was determined and is given in Fig. 9.27. One notices a prototypical saturation of D_2 at ≈ 2 which is in perfect agreement with the conclusions drawn from interpreting the return map in Fig. 9.23. Note also the excellent scaling properties as evidenced by the extremely small estimated error in D_2. However, the latter result is worth nothing without verification by surrogate data. Fig. 9.28 thus gives $D_2(d)$ for the shuffled data. As can be seen immediately, the structure in phase space was real in that it gets destroyed by the randomisation. As also observed in Sect. 9.2.1, the uncertainty in D_2 increases due to worsened scaling properties.

The above findings may be summarised as

- The data possesses some non-trivial structure (signal itself, phase and return plots)
- The attractor should be about two-dimensional (return map)
- There is evidence for chaoticity from the shape of the probability distribution
- Noise is also ruled out by the IFS-clumpiness test

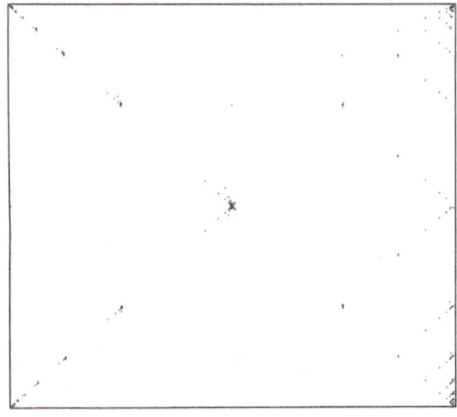

Fig. 9.25. IFS-clumpiness test for chaotic Lorenz data

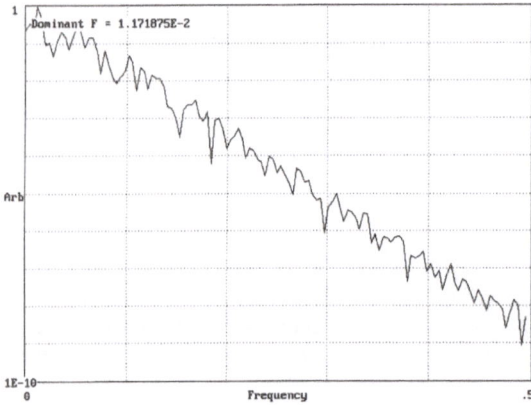

Fig. 9.26. Power spectrum of chaotic Lorenz data

– Further evidence for chaoticity comes from the power spectrum (broad, $1/f$, no dominant peaks), also ruling out quasi-periodicity
– The data possesses an attractor of dimension of about two $(D_2(d))$

The listed findings inevitably lead to the conclusion of low-dimensional chaotic dynamics which could most probably be modelled by three equations. Note that the latter conclusion agrees very well with the known properties of the dynamics and that the expected value for the Lorenz attractor is about 2.05.

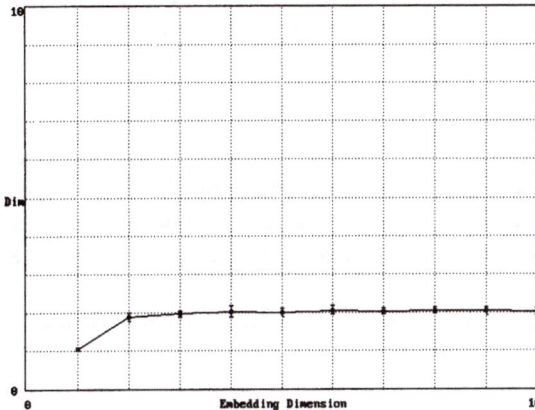

Fig. 9.27. $D_2(d)$ for $d = 1$ to 10 for chaotic Lorenz data

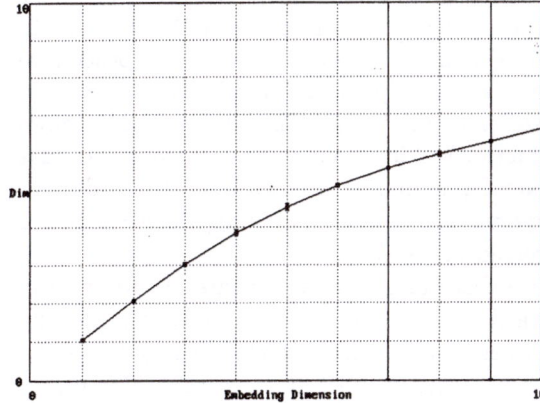

Fig. 9.28. $D_2(d)$ for $d = 1$ to 10 for shuffled chaotic Lorenz data

9.3 Analysis of Earthquake Data

In this section, geophysical field observations thought to be related and possibly describing the earthquake process or at least certain aspects of it, are analysed in the fashion of the previous section. The results obtained earlier will be helpful in the interpretation and classification of the results to follow.

9.3.1 Radon

Radon is the only radioactive gas emitted from the surface of the crust and the emission is thought to be predominantly governed by stress changes in the basement rock (e.g. [WNS88]). Changes of radon concentration in groundwater with time are considered as promising earthquake precursors (e.g. [Wak82]).

The radon concentration time series used here was kindly made available by Dr. G. Igarashi of then Laboratory for Earthquake Chemistry at University of Tokyo. The original data comprised hourly values of the raw observed radon concentration, the temperature in the measuring chamber (see below), a calculated temperature response, a calculated radon concentration in the liquid phase and the resulting (raw - temperature response) time series. The radon gas concentration is measured by a scintillation detector system in a detection chamber. Since the emanation rate of radon is proportional to the temperature in the chamber, the latter has to be monitored and taken into account to obtain the actual concentration of radon in the groundwater (see [NW77] for a detailed description of the system).

Tokyo University operates radon observation sites at several locations in Japan of which the site KSM in the eastern part of Fukushima Prefecture, Northeast Japan, has shown exceptional sensitivity to the occurrence of earthquakes ([IW90]). The geographical location of site KSM is shown in Fig. 9.29. Igarashi and Wakita (1990) also outline the statistical procedure used to reduce the raw data to the "earthquake relevant" part by removing the temperature effect and an "irregular" part. The latter already shows that smoothing (i.e. "noise"-reduction) is involved in the approach so that it is necessary to use the raw data for nonlinear analysis (recall the dimension reducing effect of smoothing and that "noise" may indeed constitute the nonlinear dynamics we are interested in here).

From the original data set, a subset of almost two years (14 401 hourly values) from Nov. 18, 1985 onwards was selected so as not to include any gaps and be stationary (the measuring procedure was changed by installing an air conditioner to reduce the temperature effect at one stage). The yearly temperature effect is clearly visible in Fig. 9.30 where the selected data is shown. The harmonic constituents are confirmed by the power spectrum as given in Fig. 9.31 (linear-linear plot).

As we are not interested in meteorological dynamics (which possess simple periodic behaviour at the available scale), the harmonic trend has to be

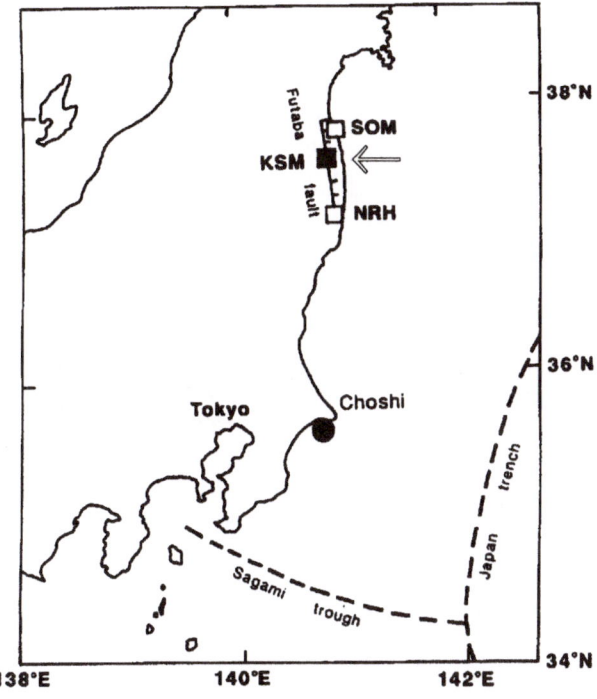

Fig. 9.29. Location of radon site KSM (after Wakita *et al.* (1991))

removed prior to nonlinear analysis. Fig. 9.32 shows an attempt to remove the slow trends by a sixth order polynomial least squares fit (the resulting polynomial coefficients are also shown in the figure). Despite the reasonable fit, dominant periodicity remained at smaller scales in the residual. The latter still dominated the phase space structure so that another approach had to be taken.

Fig. 9.33 shows the residual after all harmonic constituents have been removed by means of a maximum-entropy (or all poles) analysis (e.g. [P+92])—this analysis is similar to a Fourier analysis but represents the data in terms of a finite number of complex poles of discrete frequency. It has advantages over a FFT when trying to extract sharp peaks from records with superimposed noise. As can be seen in Fig. 9.33, the residual is very small in amplitude (note the different scales of the y-axis) and of little variability. The latter essentially implies that the data could in principle be modelled to sufficient accuracy by the superposition of harmonics alone. Thus, the analysis as outlined in Fig. 9.1 could end here (at step (C)) in principle.

As it is unsatisfying to conclude that the emission of radon gas as observed in the time series concerned here is a simple periodic process, some further analysis was performed on the above residual nevertheless. The power spectrum of the non-harmonic residual as given in Fig. 9.34 is broad and flat as was to be expected, indicative of white noise (cf. Section 9.2.2). Looking

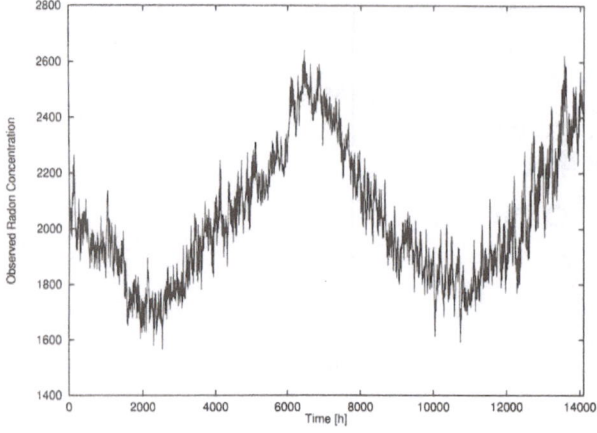

Fig. 9.30. Observed hourly radon gas concentration at KSM

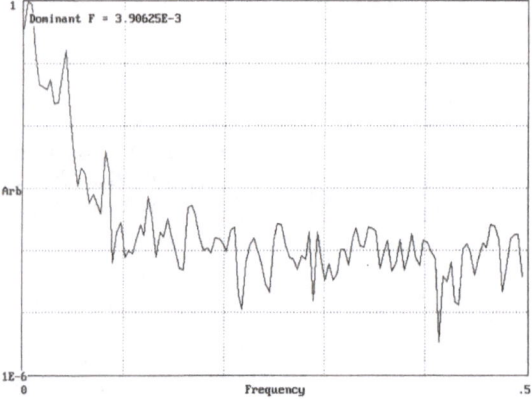

Fig. 9.31. Power spectrum of radon observation at site KSM

at the IFS produced by the IFS clumpiness test in Fig. 9.35, this assumption is confirmed (cf. Section 9.2.2): the attractor of the IFS is homogeneously space filling, no structure is apparent. The autocorrelation function, given in Fig. 9.36, however, is surprising at first sight because it does not drop to zero abruptly and does not stay there. Instead, a clearly periodic behaviour is observed which can nevertheless only be attributed to the incomplete removal or the artificial introduction of periodicity by the all poles method. Hence, one would interpret the autocorrelation function as evidence for uncorrelated noise as well.

To convincingly confirm the latter findings, a delay time embedding was attempted as in Section 9.3.3 above. The resulting $D_2(d)$ function is displayed in Fig. 9.37. There is no apparent saturation of D_2, even the scaling properties themselves are questionable for $d > 6$ as evidenced by the large error bars. Even if one would want to believe in a saturation of the attractor dimension at about 7 this would imply essentially random data in the case of the limited length of the time series regarded here (cf. [SR95]). The latter is at least true

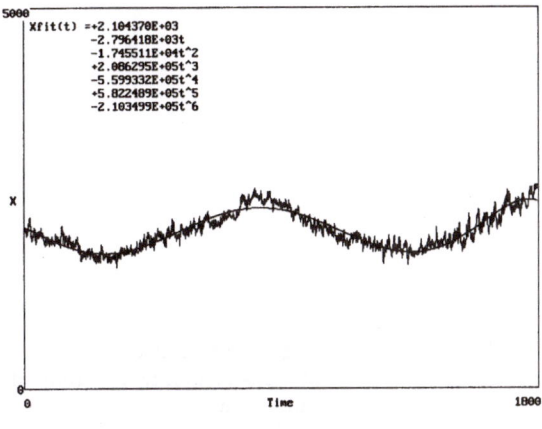

Fig. 9.32. 6th order polynomial fit to radon data to remove slow trend

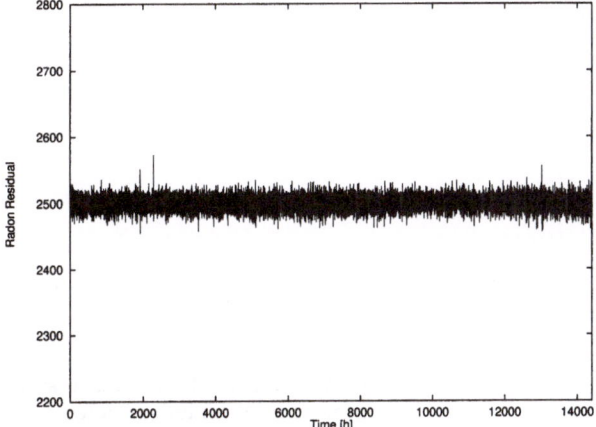

Fig. 9.33. Radon residual after a maximum-entropy analysis

for the current state of numerical modelling (the limit must be seen at about $D_2 = 5$).

A final attempt to detect low-dimensional determinism was made by looking at the radon fluctuation signal, similar to the fluctuation of earthquake inter-arrival times in Section 9.3.3. In Section 9.3.3, there was no choice but the fluctuation signal, but here signals like the ones tried above (i.e. residuals) are preferable in principle because taking the numerical derivative emphasises measurement error. Only the result of attempted delay-time embedding is given in Fig. 9.38 as the other results (phase space plot etc.) were rather similar to the ones of the detrended data. As can be seen, Fig. 9.38 may be regarded to be almost identical to Fig. 9.37: the attractors of the detrended data and the first derivative of the raw data show no apparent structure in embeddings of dimension up to 10. The embedding results confirm each other and thus also the maximum-entropy method as used for the removal of harmonic constituents of the data.

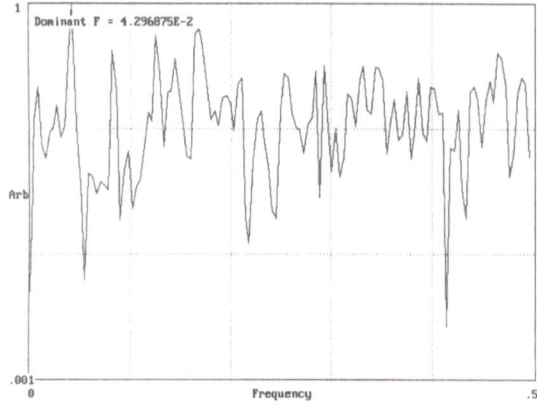

Fig. 9.34. Power spectrum of non-harmonic residual of radon data

Fig. 9.35. IFS clumpiness test of non-harmonic residual of radon data

To summarise the analysis of radon data, one may conclude that the dynamics of radon emission as witnessed here (i.e. also at the time scale of about two years) must be modelled by harmonic processes instead of low-dimensional nonlinear processes. The small non-harmonic residual cannot be regarded to be a low-dimensional nonlinear process either; it is either high-dimensional or uncorrelated noise.

To list the above results concisely, it may be said that

- The main constituent of the data is harmonic (power spectrum of raw data, small residuum)
- The residual is white noise (power spectrum of residual, IFS-test, autocorrelation)
- There is definitely no low-dimensional determinism in the non-harmonic residual ($D_2(d)$)
- The same is true for radon fluctuations ($D_2(d)$)

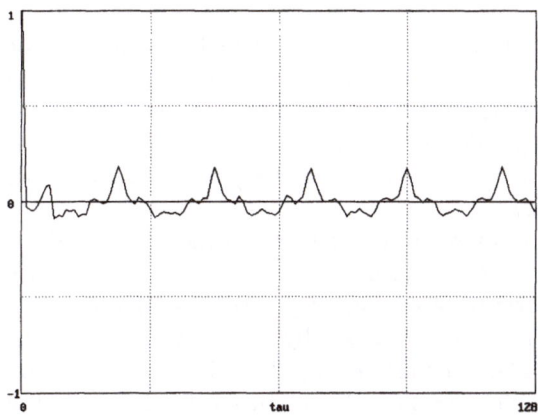

Fig. 9.36. Autocorrelation function of non-harmonic residual of radon data

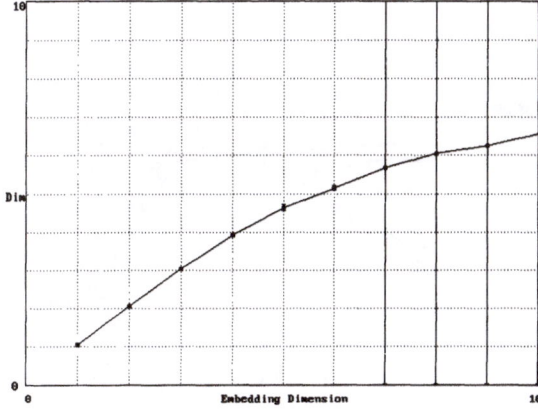

Fig. 9.37. $D_2(d)$ as obtained for delay-time embedding of non-harmonic radon residual

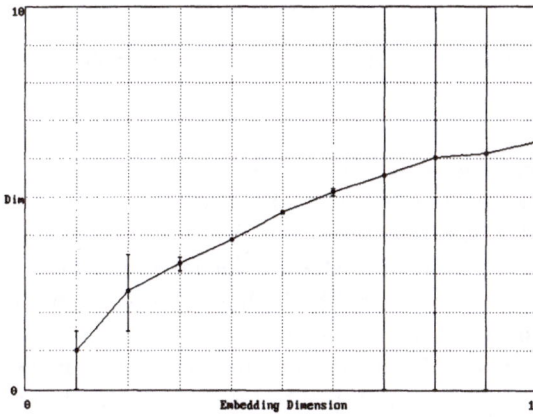

Fig. 9.38. $D_2(d)$ as obtained for delay-time embedding of radon fluctuation

9.3.2 Strain

In this section, strain data as observed by an extensometer array will be analysed for determinism. The extensometer measurements commenced in 1975 at the Yamasaki fault in Southwest Japan within the framework of the national project for earthquake prediction in Japan. Details of the project as carried out in the Yamasaki fault region and about the strain measurements may be found in [Wat91a, Wat91b].

Here, the aim is rather single-mindedly to perform a time series analysis in the light of possible nonlinear determinism instead of trying to correlate the strain signal with the occurrence of earthquakes (see [Wat91b] for such an analysis). The data used for analysis was kindly made available by Dr. Kunihiko Watanabe of DPRI. The left part of Fig. 9.39 shows the location of the Yamasaki fault system which includes the Yasutomi fault, where the measurements were performed. The geometry of the extensometer vault is shown to the right. Also shown are the locations of extensometer sensors, denoted by numbers 1 through 12.

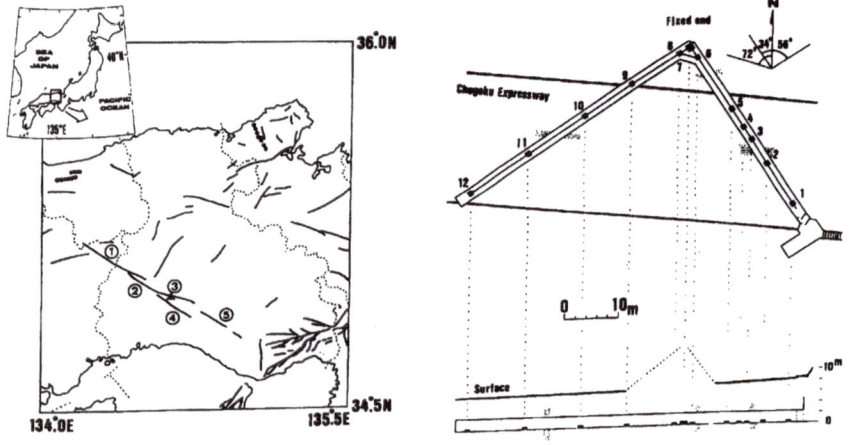

Fig. 9.39. Location and details of strain observation vault at Yamasaki fault (from Watanabe (1991))

The available data comprised time series for the deformations between sensors 1-6 and the fixed end, between points 8-12 and the fixed end and between sensors 6 and 8. Differential strains may thus be obtained by calculating the differences between appropriate sensors. The shaded elongated areas indicate fractured zones which may be regarded as "the fault"—they agree with the confirmed fault line as also observed at the surface (cf. [Wat91a]). Also in the light of the findings by Kagan (1992) mentioned in Section 4, a more detailed discussion of the term "fault" is appropriate here. As pointed out in [Wat91b], the Yasutomi fault can be characterised by a "fault zone" and a "fractured zone". The fault zone may be defined as a belt of 100-200 m thickness with several fractured zones consisting of fractured rock and clay inside. The influence of such a complicated interface on stress accumulation, strain release and rupture behaviour in general and thus also on the signal discussed here is certainly very difficult to model. In the choice of data sets for nonlinear analysis it was simply assumed that the extensometer signal across the "fault" might be particularly interesting as it can be considered to be especially sensitive (cf. Kümpel (1993) and his introduction of "weak-intersecting-zones") and to in a sense describe the thickness of the fault. The differential strains between sensors 2 and 3 and 10 and 11 were thus calculated.

The original data consisted of hourly values for Jan. 1984 to Dec. 1985 and daily values from Nov. 1975 to Dec. 1987. The differential strains were calculated for both types of data to be able to check for sampling effects respectively short term and long term behaviour. The raw data from the sensors is subjected to a low pass filter with a cut-off period of one minute at the site to remove noise caused by traffic on the expressway also indicated in Fig. 9.39. Hourly and daily data is not obtained by smoothing or averaging the data, but by simply sampling values at appropriate intervals. Thus no further low pass filter has been applied which would effectively reduce the dimensionality of the data (see remarks on filtering in Section 9.3.3). Temporal accuracy of the data is 2 min, the detectability limit is about 10^{-9} strain. The only additional pre-processing carried out here has been to remove artificial disturbances. For the analysis, strain steps, which did cause a permanent offset, have been removed because they were too few to make feasible a resolution of their dynamic behaviour. They would hence just have caused a translation of part of the attractor in the reconstructed phase space. Because of the pronounced seasonal variation, respectively a strong linear trend in the data (cf. also [Wat91a]), the strain fluctuation signal (i.e. the first numerical derivative) was used (similar to Section 9.3.1, spectral and others methods yielded no significantly different results). The long-term features should be removed definitively because the data sets are not long enough to resolve their possibly nonlinear dynamics. After preliminary numerical inspection it was decided to use the daily data from sensors 10/11 and the hourly data from sensors 2/3. The latter choice was also motivated by the goal to compare the signals originating from the two different tunnels of the observation vault and to be able to possibly resolve the dynamics at different scales and different temporal resolutions.

The resulting fluctuation time series are shown in Fig. 9.40 for the daily data and Fig. 9.41 for the hourly data. The amplitude of strain fluctuation is similar for both time series but the hourly record contains several large spikes (due to the occurrence of larger earthquakes, cf. [Wat91b]) which were missed when sampling the daily values (recall that the daily values are not produced by averaging or the like but simply by selecting values at 0:00 hours each day). Thus no fundamental difference between the time series due to their numerical preprocessing are expected. Also no fundamental differences due to the different locations of the sensors are apparent at this stage.

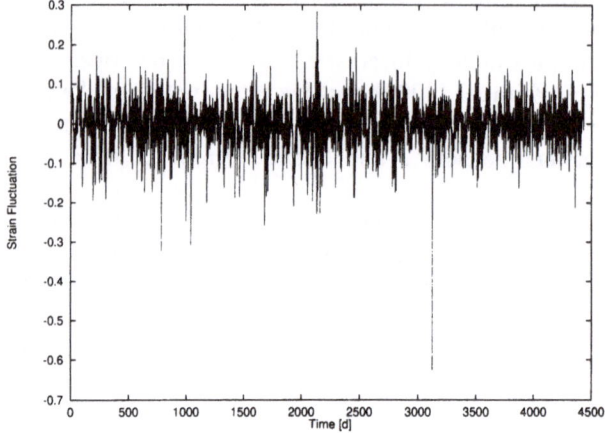

Fig. 9.40. Daily strain fluctuation at the Yamasaki fault for 1975 to 1987

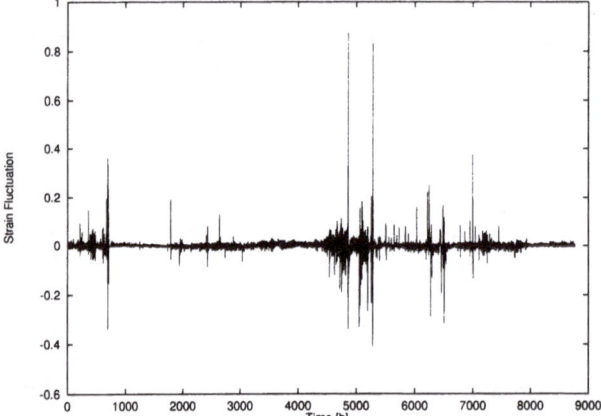

Fig. 9.41. Hourly strain fluctuation at the Yamasaki fault for 1984 to 1985

All the analyses mentioned in the previous two sections were carried out for the two sets but only a few are shown below. Due to the shape of the original data sets (i.e. before taking the first derivative) being reminiscent of a random walk (they possessed pronounced slow trends which could be described as (transient) excursions from the mean), first the Hurst exponents for the actually observed strains were calculated. H resulted to be 0.58 and 0.60 for the hourly and daily data, respectively. The latter values are very close to Brownian motion, i.e. the integral of white noise—the original data possesses no pronounced long-time memory which seems to be rather the exception than the rule with geophysical data (cf. Section 5.1).

Phase space plots and return maps unveiled no discernible fractal structure for both sets so that further judgement had to be performed by numerics. Fig. 9.42 shows a delay-time embedding in three dimensions ($x(t)$ vs. $x(t-1)$ vs. $x(t-2)$ with $\tau = 1$) to demonstrate the structure-less phase space of the strain fluctuation for the daily data: the attractor forms a dense ball without discernible (fractal) structure.

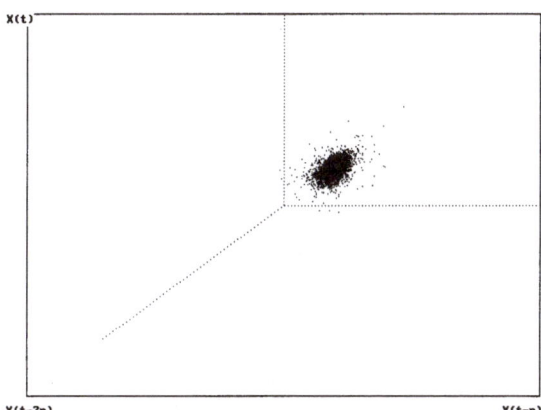

X(t)

X(t-2n)   X(t-n)

Fig. 9.42. Three-dimensional delay time embedding of daily strain fluctuations revealing no discernible structure

Results for daily and hourly strain fluctuations were very similar so that in the following only the results for the hourly data are given. The logarithmic power spectrum in Fig. 9.43 is broad but shows a dominant periodic constituent at a frequency of about $0.16\times1/2\text{h}$. There is no power law decay or other structure which might be indicative of a chaotic data.

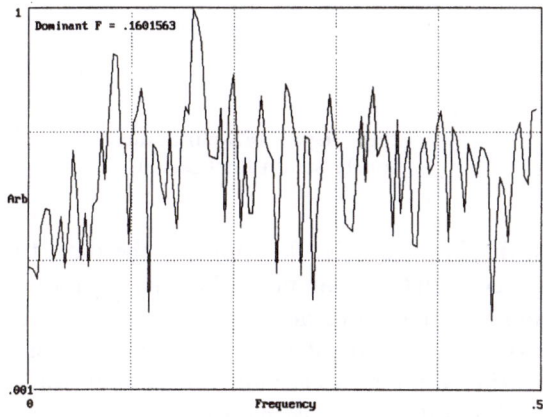

1

Dominant F = .1601563

Arb

.001
0   Frequency   .5

Fig. 9.43. Logarithmic power spectrum of hourly strain fluctuations

While the histogram (probability is plotted versus the binned range in strain fluctuation) in Fig. 9.44 is not definitely typical for uncorrelated noise (or periodic data), the IFS-clumpiness test in Fig. 9.45 seems to indicate noise: the attractor of the IFS spreads uniformly over the whole area when compared to e.g. Fig. 9.25 (Lorenz Chaos).

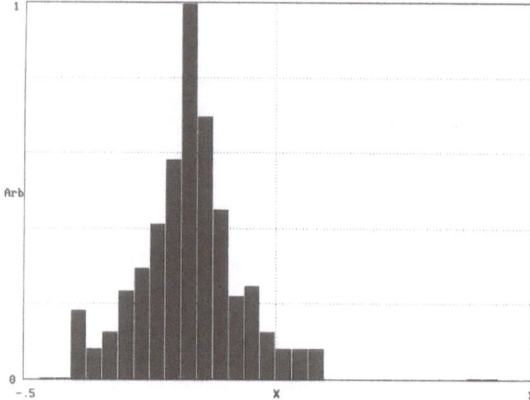

Fig. 9.44. Logarithmic histogram of hourly strain fluctuations

Fig. 9.45. IFS-clumpiness test of hourly strain fluctuations

The autocorrelation function displayed in Fig. 9.46 unveils the data to possess a somewhat higher degree of correlation than pure white noise (cf. Fig. 9.16): the coefficient of correlation can be seen to decrease more slowly and vary with time, possibly indicative of the dominant periodicity as witnessed in Fig. 9.43 above. The latter situation is similar to the autocorrelation result of the non-harmonic radon residual and does not distinguish the strain data from noise.

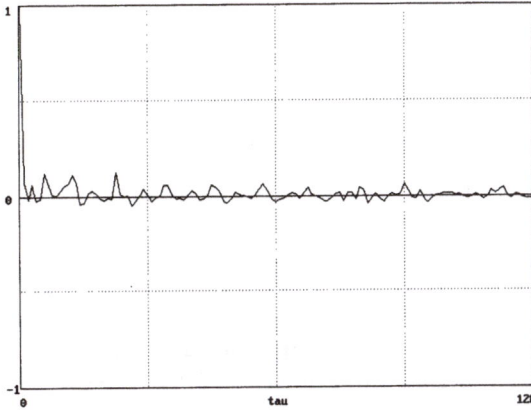

Fig. 9.46. Autocorrelation function of hourly strain fluctuations

The above results lead to the conclusion that one cannot expect to encounter a saturation of the fractal dimension of the attractor at any embedding dimension. The latter is confirmed in Figs. 9.47 for the daily data and in Fig. 9.48 for the hourly time series: for the daily data, $D_2(10)$ has surpassed the value of 6, the hourly $D_2(10)$ is only a little lower at about 5.6. Despite the fact of an apparent trend towards saturation of the attractor dimension in both figures, the maximal values are so large that deterministic modelling is out of the question. Note also the poor scaling properties as evidenced by the indicated error bars. Finally, Fig. 9.49 underlines the latter conclusions of the data being essentially noise: the figure shows $D_2(d)$ for the hourly data after it had been scrambled in the time domain. The fact that this curve shows essentially the same behaviour as for the original data shows that the latter possessed definitely no determinism.

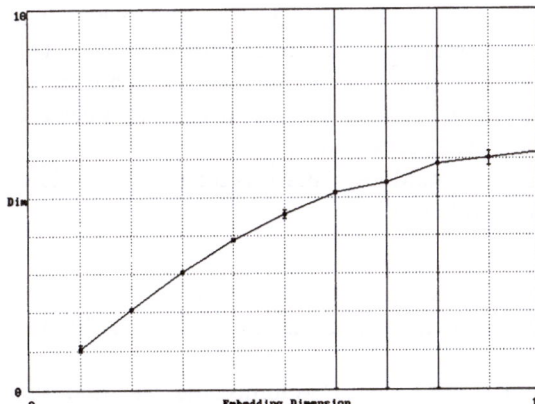

Fig. 9.47. $D_2(d)$ for embeddings of daily strain fluctuations

An advanced analysis of the strain data discussed in this section would be to consider e.g. the differential strains between all sensors simultaneously

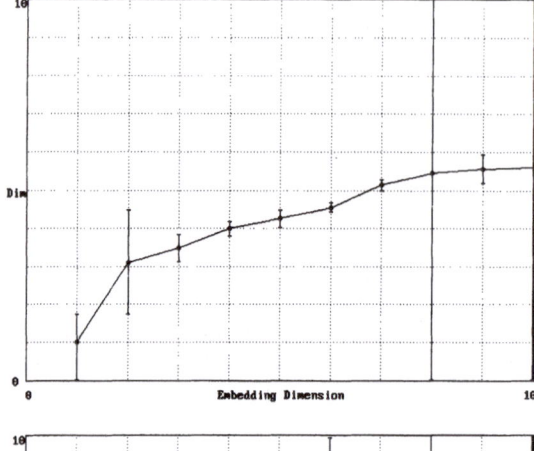

Fig. 9.48. $D_2(d)$ for embeddings of hourly strain fluctuations

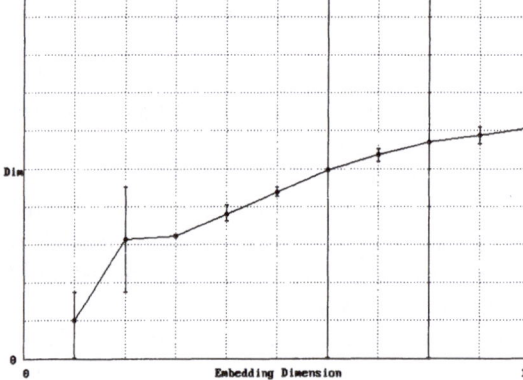

Fig. 9.49. $D_2(d)$ for embeddings of scrambled hourly strain fluctuations

(multivariate data). The latter approach would greatly enhance the information content in that not only one scalar signal is used as input but rather a strain field (cf. also [ABST93]). A further possible advantage would be the possibility of the automatic reduction of noise, as the noise components of the different sensors can be regarded to be independent (at least to a certain extent) and thus would lead to a mutual cancellation.

One may summarise the findings of nonlinear analysis for the strain data (for hourly as well as daily sets), as follows:

- The data shows very little persistence (H)
- The data possesses no structure (phase space plot)
- The data does not stem from a well-defined small number of modes (power spectrum)
- Further evidence for noise comes from the IFS-clumpiness test and the probability histogram
- The autocorrelation function also is indicative of noise
- There is no structure in higher dimensions either ($D_2(d)$)

9.3.3 Inter-arrival Times

A "time" series directly derived from the occurrence of earthquakes is analysed here. From the catalogue introduced in Chapter 7, a record of earthquake inter-arrival times, i.e. earthquake intervals, was constructed. Such a "time" series has the advantage of undoubtedly directly describing an aspect of the earthquake process. As will be discussed below, the latter is not necessarily true for geophysical time series obtained from field measurements. Obviously, the x-axis is not really time, but for convenience, it will be spoken of here as a time series. It makes no difference for analysis and interpretation (the simultaneous measurement of a signal at several different locations is equivalent to a scalar measurement for a prolonged time at only one location, cf. also [ABST93]).

Following the approach outlined in Fig. 9.1, it was first made sure that none of the 27951 values was due to artificial disturbance (e.g. gaps in the earthquake catalogue etc.). The resulting data obviously has a structure of impulses on a line, which is not feasible for viewing and analysis in phase space (a dense structure with rectangular boundaries will result). Hence, although no trend(s) or other obvious non-stationary behaviour was detectable, the signal was differentiated to give a signal describing the fluctuation of earthquake intervals. The latter signal was used for all subsequent analyses; the first 15 000 values are shown in Fig. 9.50.

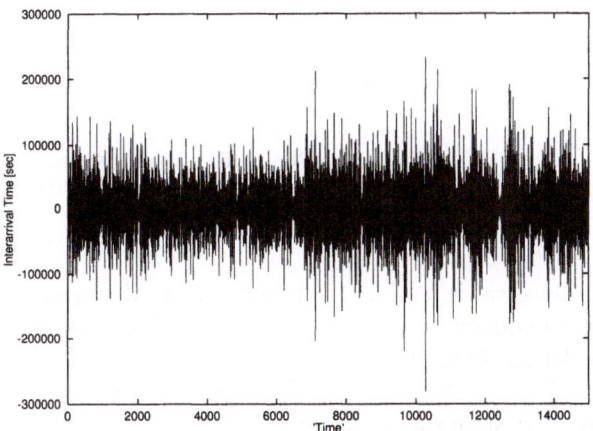

Fig. 9.50. First 15 000 values of the fluctuation of earthquake inter-arrival times

Next, step (C) involved the search for simple structure which would end the analysis if the data could e.g. be sufficiently modelled by a superposition of harmonics (cf. Section 9.2.1) or a polynomial. In the latter case, also the phase space plot would reveal a simple topography like e.g. a (noisy) torus (also cf. Section 9.2.1). As this step includes some experimentation in the case of real-world data, like varying the delay time and angle of view in phase space etc., it is difficult to reproduce the results on paper. The latter is especially true for animated sequences like stroboscopic sampling of phase space (cf. [SR95]). A few examples are nevertheless given in the figures to follow.

Fig. 9.51 shows the power spectrum of the data, which should be broad (dynamics not representable by simple superposition of harmonics) and follow an exponential or power law not to exclude the possibility of low-dimensional deterministic chaos (the latter would appear as a straight line in a log-linear plot, cf. Section 9.2.3). The spectrum was determined by applying a non-overlapping Parzen window using a FFT (cf. [P+92]). As can be seen, the power spectrum is broad but definitely doesn't represent $1/f$-noise. Instead there seem to be two linear regions, one increasing with f and one rather flat. Although one may exclude the possibility of periodicity in the data according to this result, the trend of the power spectrum is puzzling. It might possibly represent coloured noise, i.e. correlated noise with a non-flat, non-trivial power spectrum.

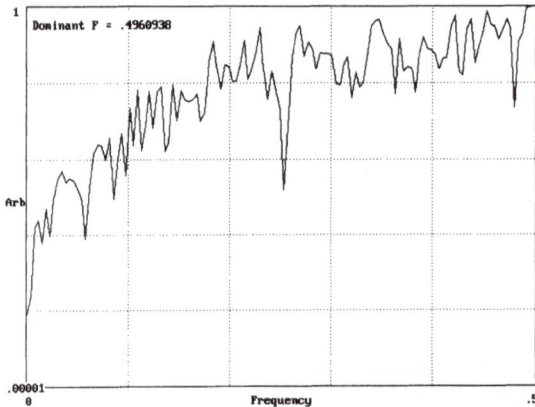

Fig. 9.51. Power spectrum of earthquake inter-arrival time fluctuation

In the case of flat power spectra, it is of interest to integrate the original data and determine the Hurst exponent as described in Chapter 5 and carried out in Section 9.2.2 above. Recall that an exponent of about 0.5 would indicate uncorrelated white noise (i.e. Brownian motion, the integral of white noise), while $H > 0.5$ means persistence and $H < 0.5$ means anti-persistence (fBm). The result of a Hurst analysis of the inter-arrival times is given in Fig. 9.52. The y-axis denotes root-mean-square displacement of the signal from its initial position using each data point as an initial condition (i.e. range), the y-axis is time. The slope thus gives H. The Hurst exponent was found to be 0.73, i.e. in accordance with an overwhelming multitude of other geophysical phenomena (see Chapter 5). It should be noted that, apart from the final divergence for large times, the plot follows the straight line exceptionally well.

Here it is sufficient to note that the occurrence times of earthquakes possess a long time memory such that current trends persist.

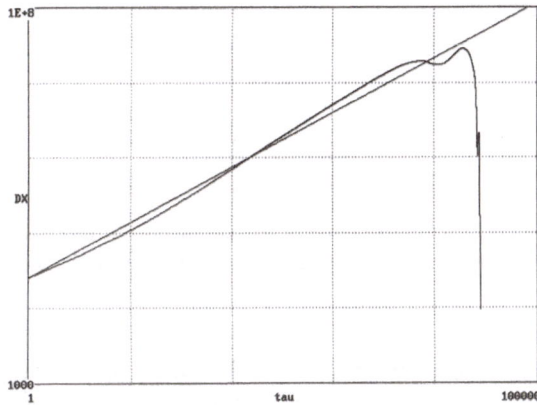

Fig. 9.52. Hurst plot of integrated earthquake inter-arrival times

A look at the probability distribution and associated analyses might bring more clarity. Fig. 9.53 gives a simple histogram of the data. The data was binned into 64 bins of equal widths and the weight of each bin, representing the probability of occurrence of the value denoted by the x-axis, was plotted. A non-fractal distribution is observed. However, the histogram is not necessarily Gaussian either; as also with chaotic systems with simple distribution of values, one can not conclude much from this result.

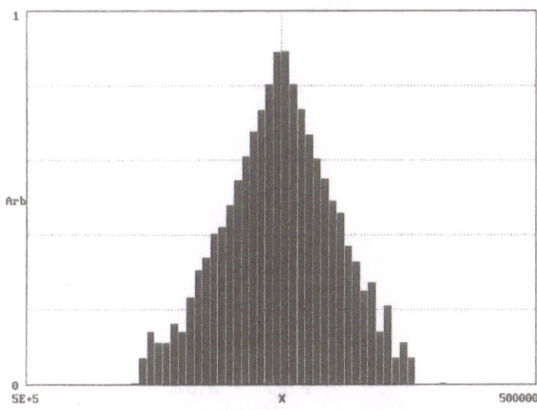

Fig. 9.53. Probability distribution of earthquake inter-arrival time fluctuation

As can be seen from the IFS-clumpiness test displayed in Fig. 9.54, however, the earthquake data definitely can not be assumed to be random which confirms the result of H and agrees with the non-flat shape of the power spectrum.

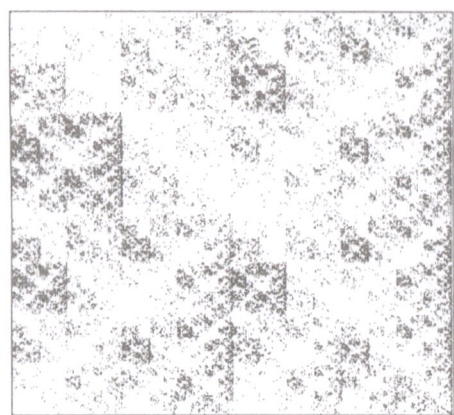

Fig. 9.54. IFS-clumpiness test for earthquake inter-arrival time fluctuation

Related to the FFT which was used above to obtain the power spectrum is the discrete wavelet transform (DWT, e.g. [P+92]). One of the main differences is that the basis functions are not imposed as in the FFT (sines and cosines) but can be adjusted to the data. Also they are localised in space (unlike harmonic functions) and frequency (like harmonic functions). One may thus obtain information about the variability of data at specific scales. Fig. 9.55 shows a DWT of the inter-arrival time fluctuation for a transform function which is symmetric about the position T where it is applied. The width ΔT of the function is plotted on the y-axis, the x-axis represents position T. As can be seen, variation within the data occurs mainly on time scales of 1 to about 8 events. As there is no systematic variation with T, we have further evidence for the stationarity of the data (cf. [SR95]).

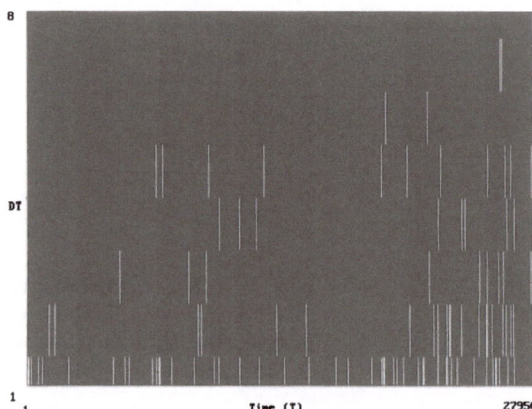

Fig. 9.55. Symmetric wavelet transform of earthquake inter-arrival time fluctuation

Motivated by the findings so far, an attempt was made to visualise a possible attractor. For this, several embeddings and phase space plots were produced. Fig. 9.56 shows an embedding of the signal into a two-dimensional phase space. Each value $x(t)$ was simply plotted versus the previous value $x(t-1)$. The latter corresponds to a two-dimensional embedding with delay time $\tau = 1$ (cf. Section 9.2). Fig. 9.56 gives a stereographic projection in which one may indeed identify some "strange" structure: random data produces a uniform shape distributed over the whole phase space (cf. Section 9.2.2) which is very different from what is observed here.

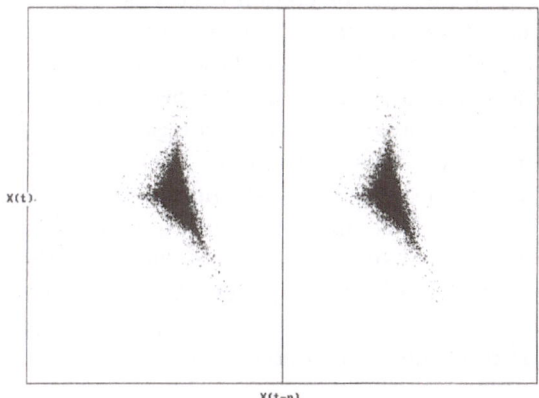

X(t).

X(t-n)

Fig. 9.56. Stereographic view of two-dimensional embedding of inter-arrival time fluctuation

Encouraged by findings like the one in Fig. 9.56, return maps were also tried. In Fig. 9.57, $x(t)$ is plotted at positions where $x'(t) = f$ (y-axis) versus the previous time where the condition was fulfilled (x-axis). Here, $f = 0.55$ which yielded the best result. The object obtained this way seems to indeed possess a fractal structure with a fractal dimension. Unlike in Section 9.2.3 (Lorenz attractor), one can not easily judge the dimensionality of the attractor. The latter will be achieved by determination of D_2 below.

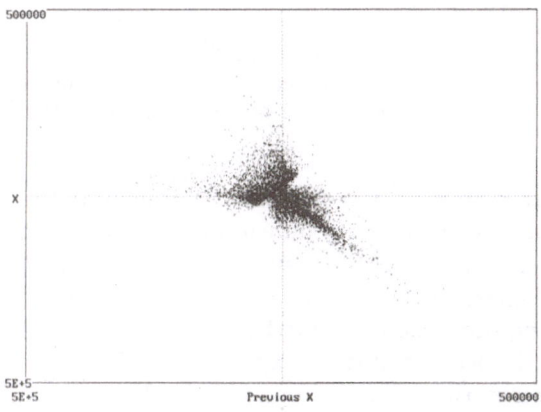

500000

X

5E+5
5E+5   Previous X   500000

Fig. 9.57. A return map of earthquake inter-arrival time fluctuation

Before steps (D) and (E), i.e. the verification and characterisation of the strange attractor, one needs to guess the optimal delay time for the embedding in successively higher Euclidean dimensions (in Section 9.2, the delay time was always assumed to be one). Several methods exist to achieve this (e.g. [ABST93]), the simplest of which is based on the autocorrelation function (e.g. [P+92]): the time τ, where the coefficient of autocorrelation first reaches $1/e$ is usually said to be the correlation time. To disentangle the structure in embedding space, a delay time greater than the correlation time is obviously desirable (e.g. [Sch88]). Here, the correlation time was 0.426 as displayed in Fig. 9.58, hence a delay time of 1 should also be sufficient. For highly random data, there is no correlation and the correlation function will drop abruptly to zero, showing the small correlation time (cf. Section 9.2.2). Completely correlated data like a sine wave will have a correlation function that varies with τ but whose amplitude is constant (cf. Section 9.2.1). Chaotic data from natural systems may produce both behaviours, depending also on the interval at which the signal is sampled (a very high sampling rate produces several consecutive points of similar value). In this case, the autocorrelation function drops to zero relatively fast, then shows negative correlation and finally returns and stays at zero. As the possibility of white noise has been excluded above, the result does not contradict the assumption of low-dimensional chaos.

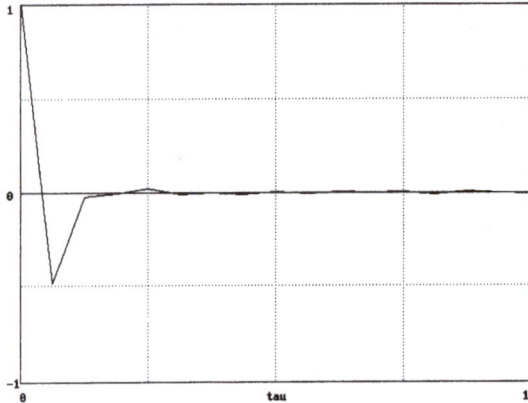

Fig. 9.58. Autocorrelation function for earthquake inter-arrival time fluctuation

Using the thus obtained delay time of 1, step (D) was performed for embedding dimensions ranging from 1 to 10. For every dimension, an embedding with delay time 1 was performed and D_2 for the resulting structure determined. If there is really a low-dimensional attractor in the data, D_2 must saturate at the nearby embedding dimension, i.e. further increase of d may not cite a further increase in D_2. Hence, a plateau is expected in a D_2 versus d plot. Simply speaking, the latter must be fulfilled because the dynamics of a deterministic system must unfold at a certain embedding dimension (the trajectories may not intersect anymore; the assumed system is deterministic, i.e. the systems state must be unambiguous), where addition of further degrees of freedom does not add any additional information. So to say, if we observe the motion of a beetle in the plane, we learn nothing new if we plot its trajectory in three or more dimensions instead of two (cf. also Section 9.1). Obviously, however, all the precautions mentioned in Chapter 2.1 must be taken (i.e. sufficient linearity of the scaling region, sufficient size of scaling range etc.). Also the mentioned limitations with respect to the needed number of data points apply—recall that an increasing dimension requires geometrically more points. The error for D_2 was estimated to be half the difference between maximal and minimal pointwise slope over a fixed scaling region. If this error remains small, the determination of D_2 may be regarded to be successful.

Fig. 9.59 gives the result of this kind of time consuming calculation. It is exciting to observe a very clear saturation at $D_2 \approx 3$. The small error in D_2 as designated by the error bars gives high confidence in the result. The latter outcome is exciting because almost all analyses of real world field data (and of most laboratory data) end at this point[3]. Compare also Fig. 9.27 in Section 9.2.3 where $D_2(d)$ is given for the noise-free Lorenz system and note that the error bars are only slightly larger despite the fact that the data here represents un-filtered (un-smoothed) real-world data!

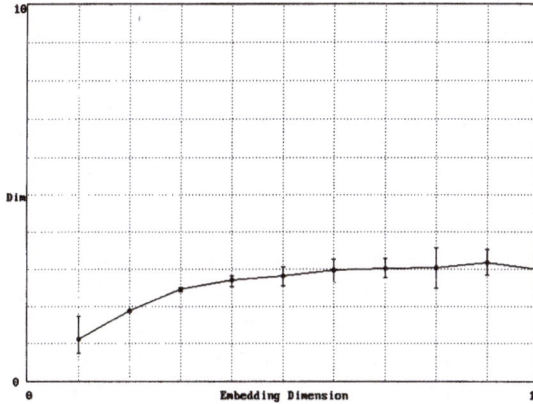

Fig. 9.59. $D_2(d)$ for embedding dimensions of 1 to 10 for earthquake inter-arrival time fluctuations

A saturation at about 3 is one dimension higher than the classical Lorenz attractor, which, recall Section 9.2.3, can be modelled by 3 coupled nonlinear differential equations. One would thus expect that the temporal occurrence of earthquakes could possibly be modelled by as few as four equations—in

---

[3] After analysing many earthquake related and other time series, including such which had been published as being chaotic, I asked Prof. C. Sprott([Spr93]), whether he knew a convincing example of numerically shown low dimensionality in real-world data and he said he knew none (see also Watts *et al.* (1994))

other words, only four variables are needed: the process has only four degrees of freedom. Turcotte (1992) reviews the Lorenz model in view of thermal convection in the earth's mantle as the driving force of plate tectonics. Because of plate tectonics being in turn the motor for earthquakes, one might be led to assume a deeper connection between the Lorenz model and the results obtained above. Should the result confirm chaotic mantle convection and the chaoticity of earthquakes? Could the occurrence of earthquakes really be modelled by such a small number of governing equations?

Obviously, analysis of many more data sets would be necessary to gain evidence for such a fact, but, more at hand, the saturation of D_2 must be confirmed to be due to low dimensional determinism first (step (F)). For that purpose, the method of surrogate data ([CE91, The91, TEL+92, KI92]) has been devised about six years ago: in this approach, the original data is either randomised in the time domain ("shuffled") or Fourier-transformed, phase-randomised and then re-transformed into the time domain. The first method is rather brutal and only preserves the probability distribution (the power spectrum and the autocorrelation function are destroyed. This method has been employed throughout Section 9.2). The second method preserves the power spectrum and autocorrelation function and only alters the probability distribution. The important fact, however, is that both methods necessarily remove any determinism in the data: the attractor is destroyed (some kinds of coloured noise, however, can survive the phase randomisation). Hence, if the results of step (D) and (E) are still the same after randomisation, the supposed attractor was spurious, the data does not stem from a (low-dimensional) deterministic system and can not be modelled accordingly.

Fig. 9.60 shows the first 15 000 values of the phase-randomised data (above) and the shuffled data (below). The phase-randomised data appears smoothed which is not surprising as a finite number of harmonics can not accurately represent noisy or chaotic data. In fact, should the systems dynamics be known to be appropriately representable by a Fourier expansion, the Fourier transform may be used to reduce the noise in the measurement[4]. Should it be known that the raw signal consists of a linear superposition of harmonics and chaotic data, the harmonics may of course be removed this way to obtain the actual signal of interest—then, characteristic of nonlinear time series analysis, the "noise" becomes the signal of interest and the linear part gets discarded. Simple examples of the latter situation are signals which have an annual trend due to seasonal temperature variation imposed

---

[4] The issue of noise reduction prior to nonlinear analysis has naturally received much attention. Besides the fact that dynamic noise (as opposed to additive measurement error noise) cannot be separated from the signal in principle; it is generally not a good idea to attempt to remove "noise" from signals of unknown systems—one might merely reduce the dimensionality of the system and thus get meaningless results in the end (cf. [TE93]). A priori noise reduction is hence not considered a valid option here and will not be discussed any further.

(see below) or signals influenced by the earth's tides. Here, such an attempt would merely reduce the dimensionality of the data.

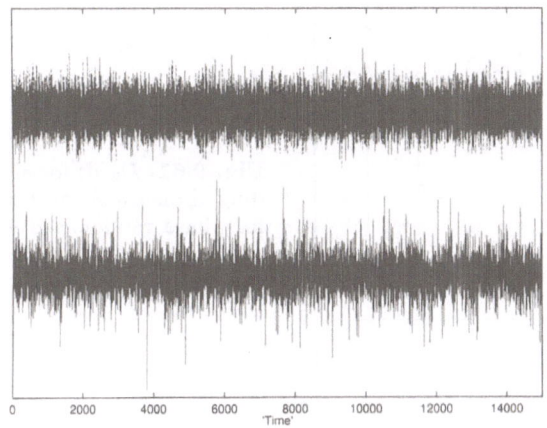

Fig. 9.60. Surrogate earthquake inter-arrival time fluctuation data sets: Phase randomised (above) and shuffled in the time domain (below)

Finally, Figs. 9.61 and 9.62 give $D_2(d)$ for the randomised data sets. One notices a very clear difference with respect to Fig. 9.59 in that no real saturation of D_2 with increasing embedding dimension may be observed any more—even the weaker phase randomisation already completely destroys the attractor. Usually, a repeated analysis of several random realisations is necessary to completely confirm the significance of the difference between the result of the original data and the random sets (by obtaining a range for D_2 from the many random sets and checking whether the original result lies within that range) but here the difference is so great, that one may be confident without such a test. One may thus exclude the possibility that the data is in fact coloured noise and we are led to believe that the temporal occurrence of earthquakes might really follow low dimensional, possibly chaotic determinism.

Before an attempt is made to characterise the degree of assumed chaos, and therefore the predictability horizon, it must still be asked whether the resulting attractor might not represent a limit cycle (some kind of periodic system) with superimposed noise. The uncertainty in the value of D_2 includes the integral value of 3—an integer dimension would exclude the possibility of chaotic dynamics (attractor is not strange). In the latter case, the motion would occur on a 3-torus. That possibility, however, can be excluded by noting the broad power spectrum, the phase space plot and the return map and by realising that it is most unlikely that the occurrence of earthquakes should be a linear process.

Additional confirmation might be obtained from the method of nonlinear prediction (e.g. [WBPP93]) which tests for noise versus chaos versus linearity by directly testing the predictability of the data (a chaotic system still

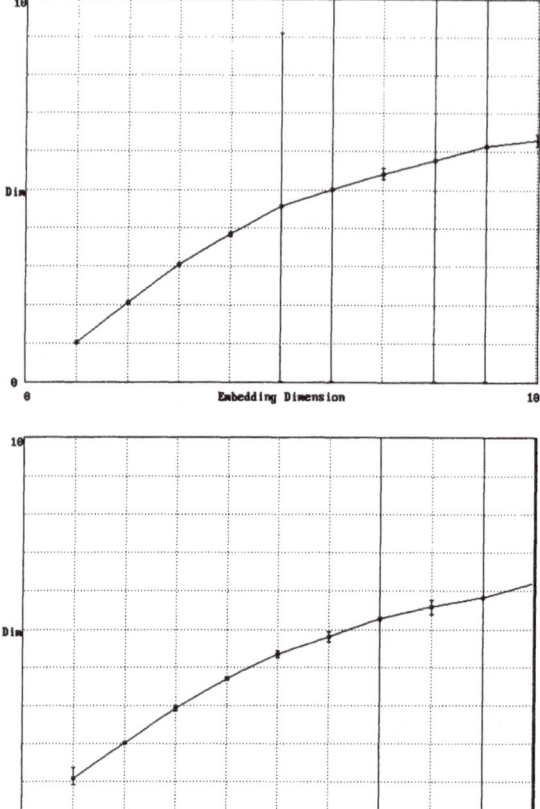

Fig. 9.61. $D_2(d)$ for embedding dimensions of 1 to 10 for phase randomised earthquake inter-arrival time fluctuations

Fig. 9.62. $D_2(d)$ for embedding dimensions of 1 to 10 for shuffled earthquake inter-arrival time fluctuations

possesses better predictability than noise, a linear system is completely predictable for all times; cf. also Section 9.2.1). As will be seen in Section 9.3.4, however, a strict confirmation of chaos is not necessary to utilise the result in the context of detection of possible earthquake precursors.

Naturally, the question arises how strong the chaoticity of the earthquake process might be. The Lyapunov exponent is a measure of the rate at which nearby trajectories of the attractor diverge (e.g. [EKRC86, SM88]). Chaotic systems have at least one positive Lyapunov exponent. For periodic orbits, all Lyapunov exponents are negative. In general, there are as many exponents as there are degrees of freedom, i.e. directions in phase space. For the inter-arrival data, only the largest exponent was calculated (cf. [WSSV85], calculation of the whole spectrum of exponents from small data sets is considered to yield extremely unreliable results). It is given in units of bits per data sample (cf. Section 2.2.4, a value of +1 thus means that the separation of neighbouring orbits doubles on the average in the sampling time). The resulting value was 0.602±0.013 for the earthquake data. Several authors have used the numerical result of such a large positive exponent as proof for chaos in the analysed data. Because the Lyapunov exponent describes the exponential divergence of solutions for nearby initial conditions, i.e. the "chaoticity" of the data, it is a measure for the predictability of the system. To be able to directly estimate the time over which meaningful predictions are possible however, the sum of all positive Lyapunov exponents has to be known. The predictability horizon is roughly of the order of the inverse of the sum of positive Lyapunov exponents to the base e. Thus, here an estimate of the time over which a prediction of the occurrence time of earthquakes to follow is not possible (though one might obtain a crude estimate for a lower bound by only using the largest exponent). Sadovskii and Pisarenko (1993) discuss the predictability of earthquakes in the light of possible seismic process phase space reconstruction and point out that the problem of how to reconstruct a seismicity attractor (for predictions) has not yet been addressed. The value of ≈ 0.6 given above would also gain increased significance if comparing different data sets, possibly also moving in time.

To sum up the results in the fashion of Section 9.2, one may say that

- The data is not (quasi-)periodic (autocorrelation, power spectrum)
- The data is not uncorrelated noise (power spectrum, H, IFS-test)
- The data is stationary (DWT)
- The data might possess a strange attractor (phase space plot, return map)
- The data is a good candidate for chaos (autocorrelation, power spectrum)
- The system represents low-dimensional dynamics ($D_2(d)$)
- The attractor must be strange (phase space plot, return map in conjunction with $D_2(d)$)

9.3.4 Monitoring Seismicity in Phase Space

The results of Section 9.3.3 led to the idea of "phase space monitoring", i.e. the continuous observation of the attractors dimensionality with time to look for possible precursory behaviour. Such a precursor would manifest itself in a gradual or abrupt change of attractor dimension, i.e. a change in complexity of the underlying dynamics which, in this case, govern the temporal occurence of earthquakes. A change like this might be indicative of a phase transition (cf. Chapter 6). A change in attractor dimension is a physically significant occurence as opposed to the mere appearance of purely statistically determined "peaks" in scalar time series. For a summary of this section see also Goltz (1997).

The same earthquake catalogue data, as in Section 7.1, was used. Figure 9.63 gives the location and the earthquake epicenters and magnitude in detail.

Figure 9.64 shows the raw time series of earthquake intervals as directly derived from the catalogue (upper part) as well as the first numerical derivative (lower part; refer to Section 9.3.3 for the motivation for taking the first derivative). Also shown in Fig. 9.67 is a delay time embedding of the same

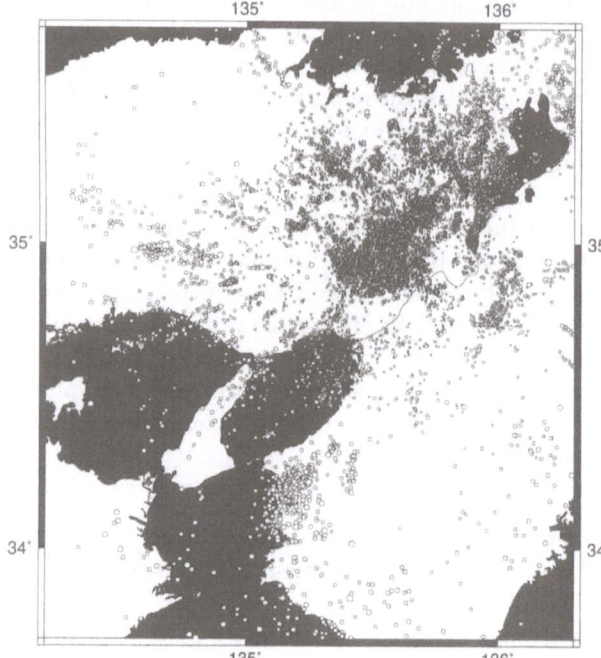

Fig. 9.63. The data used for monitoring in phase space. Shown are epicenters and magnitudes (circle diameter) of all events contained in the catalogue analysed

data with $\tau = 1$ in two dimensions to give an idea of the embedding process. An embedding into two dimensions corresponds to the first step only when looking for a deterministic structure in phase space by testing for saturation in the fractal dimension of the phase space structure with increasing embedding dimension.

To demonstrate the difference between earthquake data and synthetic data, Figs. 9.65 and 9.66 show the same information as Fig. 9.64 for Gaussian noise and Poisson noise. Recalling that Poisson distributions are used to model events occuring independently in time (such as radioactive decay, see e.g. [P+92]) and are still used to model the occurrence of earthquakes in some contexts, special attention is paid to the comparison between the earthquake data and the Poisson data.

As has been pointed out earlier, a comparison between the scalar representations of the time series is not very profitable. Looking at the two dimensional embeddings in Figs. 9.67(earthquakes), 9.68(Gauss) and 9.69(Poisson) however, some significant differences become apparent already. The shapes of the attractors differ clearly between the earthquake data and the Gaussian noise. The probability density distribution (probability is shown in shades of gray and increases towards the center in each case) of the earthquake data and that of the Poisson data also disagree. The "spiky" shapes of earthquake and Poisson data result from the occurence of rare events in the sense of extremely short respectively extremely long intervals between events followed

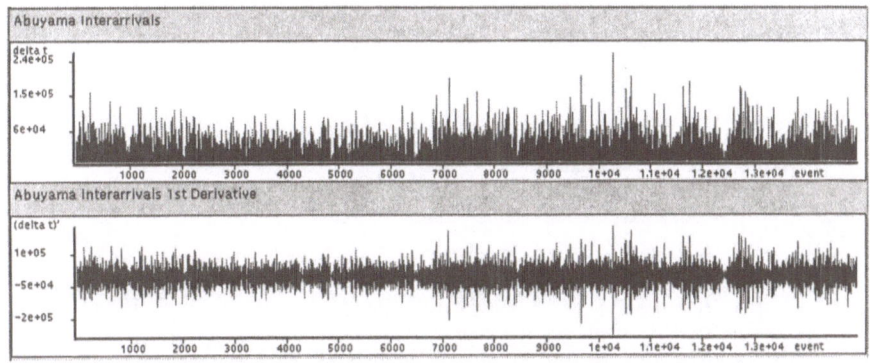

Fig. 9.64. Earthquake inter-arrival times and first derivative showing the fluctuation of lengths of time intervals between successive events

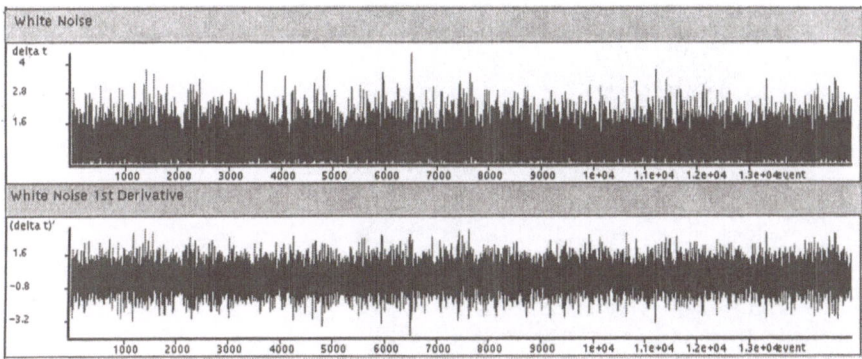

Fig. 9.65. Gaussian inter-arrival times and first derivative showing the fluctuation of lengths of time intervals between successive events

by either extremely long, extremely short or average intervals (three spikes, embedded is not the time series of intervals but its first derivative).

To clearly establish the apparent difference between these three time series, respectively their underlying dynamics, the embeddings have of course to be carried out in successively higher dimensions. The result of this complete analysis is given in Fig. 9.70. Shown are, in the order of listing in the legend, the curves for

a overall seismicity of the area of observation (cf. Fig. 9.59)
b surrogate data of the above (cf. 9.61) to exclude spuriousness
c Kobe aftershocks
d surrogate data of aftershocks
e Poisson data
f Gauss data
g Lorenz system (cf. 9.27) for comparison

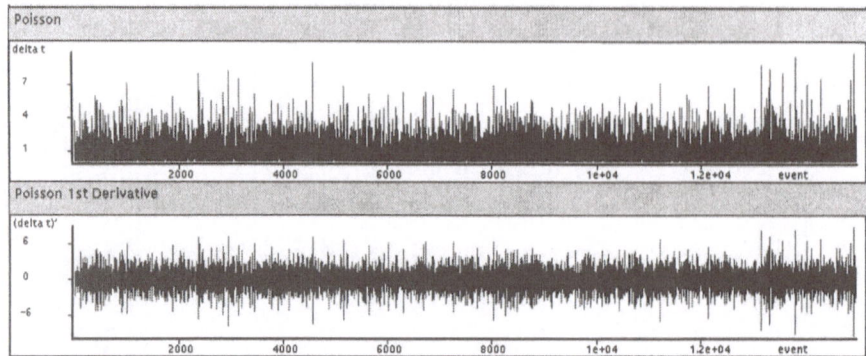

Fig. 9.66. Poisson inter-arrival times and first derivative showing the fluctuation of lengths of time intervals between successive events

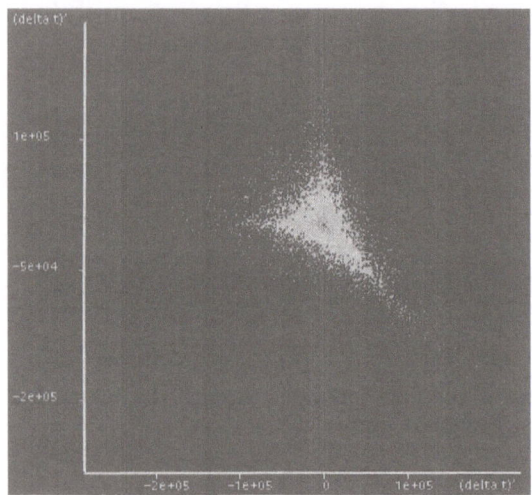

Fig. 9.67. Two dimensional phase space embedding with $\tau = 1$ of earthquake inter-arrival data

The first two curves simply confirm the results of Section 9.3.3. Analysis of the aftershocks of the Kobe earthquake yields the remarkable finding that the aftershock sequence is also low-dimensional and more importantly, that the complexity of aftershock dynamics is higher than that of overall seismicity. The latter indicates that attractor dimension really is able to distinguish different phases of seismicity and might therefore be suitable for precursor detection. Analysis of the randomised aftershock sequence confirms the existence of determinism in the original aftershock data.

Another very important result is that there is absolutely no difference between Gauss, Poisson and the surrogate data sets! Thus, what was already anticipated from inspecting the two dimensional embeddings above has been confirmed: the temporal occurrence of earthquakes is not random and cannot be modelled by a Poisson sequence.

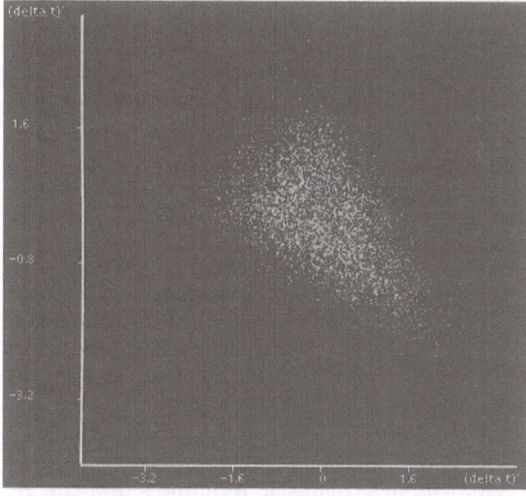

Fig. 9.68. Two dimensional phase space embedding with $\tau = 1$ of Gauss data

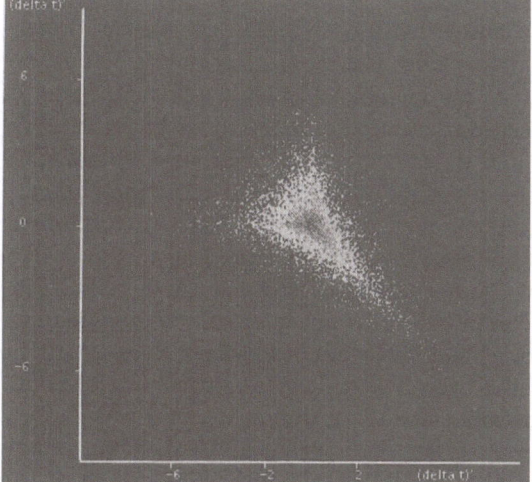

Fig. 9.69. Two dimensional phase space embedding with $\tau = 1$ of Poisson data

Because of the lack of an absolute proof of low dimensional chaos in earthquake intervals, mainly due to the difficulties in determining D_2 (cf. Chapters 2.1 and 3), the term "complexity" is preferred to "attractor dimension" in general in the following. Complexity is used in the sense of an effective degree of freedom or, in other words, regularity respectively irregularity of behaviour. Dimension is used in the sense of an effective dimension whose relative differences are significant. This convention seems appropriate in the context of precursor detection as one is not trying to proof chaoticity or to actually model the process. The same approach has been taken by Lehnertz and Elger (1998) who look for nonlinear precursors of epileptic seizure in EEG recordings. They say that "...instead of using D_2^{eff} as an absolute measure

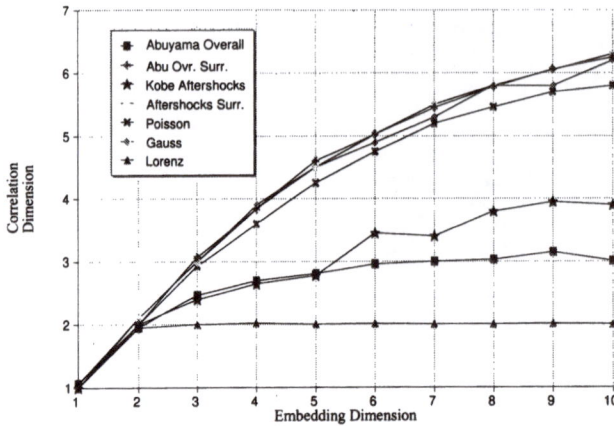

Fig. 9.70. Correlation versus embedding dimension for several time series (see text and legend in picture)

to differentiate between periodic, chaotic, or stochastic dynamics one can regard D_2^{eff} as an operational definition and use the term 'dimension' in an informal sense". D_2^{eff} denotes a numerically determined correlation dimension as opposed to a possibly existing absolute theoretical one. The authors further state that they use D_2^{eff} as an operational measure of complexity of the EEG and find that "a reduced dimensional complexity of brain activity, as soon as it is of sufficient size and duration, can be regarded a specific feature defining states which proceed a seizure". Traditional linear methods are mentioned by the authors to produce much inferior precursors at the most.

The above results for the earthquake data are encouraging to try a moving analysis of complexity throughout the 19 years of the earthquake catalogue used. In this attempt of "monitoring in phase space" the embedding dimension for all temporal windows has to be kept constant. Because of the saturation of D_2 at about 3 for the overall seismicity and the value of about 4 for the aftershock data, a constant embedding dimension of 4 was chosen. Furthermore, the window size was set to 2000 events and individual analyses were carried out with an overlap of 100 events each. The choice of the latter parameters is somewhat arbitrary but tries to balance high temporal resolution with the requirement of enough points for a safe determination of fractal dimension. The result is given in Fig. 9.71 together with an equivalent analysis of the Poisson series (upper curve). The temporal unit was converted to days instead of event number to be able to directly judge the duration of effects. Distance between successively plotted momentaneous states of complexity are thus not equidistant. Values are plotted at the time the respective windows end.

Starting with the Poisson series, one observes no significant fluctuations of complexity beyond the error bar[5]. Instead, the dimension stays at about

---

[5] The error bars are estimated from half the difference between the minimum and maximum local slopes over the middle one quarter in the correlation sum in the

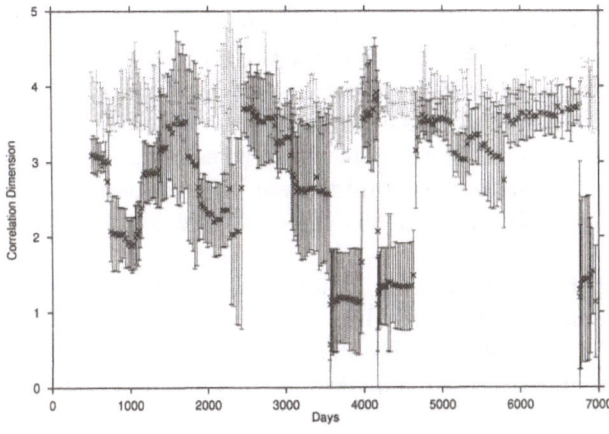

Fig. 9.71. Attractor dimension versus time in days for Poisson data (upper curve) and earthquake data

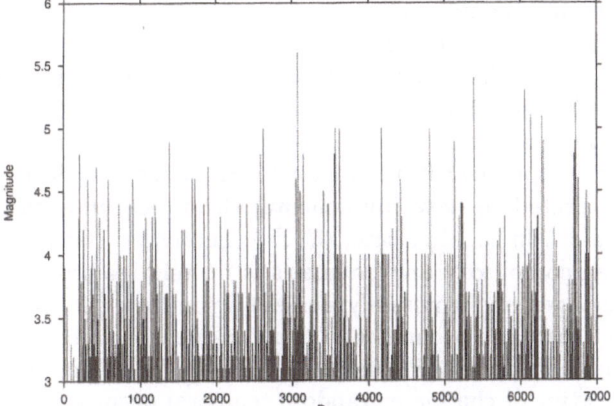

Fig. 9.72. Magnitude versus time of the earthquake catalogue analysed

4, i.e. the attractor completely fills its embedding space, its states are homogeneously spread out, the hallmark of noise. The result perfectly agrees with Fig. 9.70 when one reads off the complexity of a little below 4 for embedding dimension 4. The picture looks quite different for the earthquake data: very clear discontinuities in complexity occur with time. These jumps are of a magnitude much larger than the estimated error in the determination of D_2. The dimension stays constant during several intervals of time (sometimes years!), occasionally showing a little variation or a weak trend, then jumps to a considerably higher or lower value.

Figure 9.72 gives earthquake magnitude versus time for the same observation interval, thus making possible a first attempt of visual correlation between the behaviour of attractor dimension and the occurence of large earthquakes. Taking only the three earthquakes with a magnitude greater

---

$\log - \log$ plots of correlation sum versus sphere radii(cf. [SR95]) and are thus quite pessimistic.

than 5, one might be led to think that complexity increases before large events. This conclusion, however, seems to be contradicted by the fact that a sharp decrease happens before the Kobe event itself, which occurs at the very end of the regarded time interval. It is obvious that the relation between changes in attractor dimension and the occurence of large earthquakes needs clarification. Other ways than the one attempted here have probably to be found to correlate the two signals and establish the quality of complexity as a precursor.

Geometrically, a higher fractal dimension indicates a more homogeneous distribution of points in embedding space while a lower correlation dimension implies a higher degree of clustering. Stronger clustering in turn means prevalent occurrence of evenly spaced earthquakes in time, i.e. no big gradients in earthquake inter-arrival times. A more homogeneous distribution of points in phase space means that a considerable number of irregularily spaced earthquakes occur. Incidentally, the parameter seismicity rate and temporal pattern has been shown by Shaw *et al.* (1992) to be the only promising precursor when trying to predict the occurence of large earthquakes during simulation with a Burridge-Knopoff model.

The observed changes in complexity (attractor dimension) might be regarded useful as an empirical approach within practical earthquake prediction research or might be interpreted to have fundamental theoretical implications for earthquake dynamics. In the first case, the momentaneous effective dimensions are regarded as complexity and emphasis is on relative temporal changes only. These changes might possibly possess precursory qualities (cf. [LE98]). In the second case, the occurence of, e.g., transitions from extremely low attractor dimension (indicative of almost periodic behaviour) to high dimension (≥ 4; indicative of chaotic or random temporal occurence of earthquakes) might be regarded as phase transitions, the hallmark of critical point phenomena (cf. e.g. [Mai96]). The existence of periodic and irregular temporal windows in a specific seismically active region would imply that the degree of predictability itself changes with time.

Obviously, further research is needed to confirm these findings. Foremost, the methods outlined have to be applied to additional earthquake catalogues of different regions, other parameters than inter-arrival times should be tried and the location of the analysed region should be varied within the catalogue coverage. An application to synthetic earthquake data produced by a time dependent load-transfer cellular automaton using a hierarchical load transfer rule (cf. [GMP98]) is currently under way. To make the approach more useful for earthquake prediction research is required into the correlation between changes in complexity and the occurence of large events. Finally, complementary and more advanced methods of nonlinear time series analysis must be employed for further confirmation of the findings.

A. A Modified Box-Counting Algorithm

Although box-counting has several disadvantages, most of them are only encountered in higher dimensions (e.g. [GWSP82, vdWS88]). In the following, a modified box-counting algorithm (Liebovitch and Toth (1989)) which has a major advantage over the Grassberger-Procaccia algorithm with respect to the determination of the appropriate scaling region is introduced.

In its form modified by Sarraille and DiFalco [Sar92, Sar93], the algorithm simultaneously produces estimates of D_0, D_1 and D_2, i.e. the first three generalised dimensions. The latter yields an additional criterion for the verification of results as $D_n \overset{!}{>} D_m$ for $n < m$.

As mentioned in Section 2.2.5, the most popular measure for the estimation of a fractal dimension is the correlation dimension D_2 which is usually determined by sphere counting. In practice, the slope of $\log r$ versus $\log C(r)$ and thus D_2 itself is usually not a constant but varies with r. Excluding multiscaling, reasons for this behaviour are

– finite size of the data set, which leads to too low a resolution for $r \to r_{max}$
– noise for $r \to r_{min}$ (usually, noise will become effective in the counts at small r only)
– underestimation of $C(r)$ near the boundaries of the set

The last item points out a general advantage of the box-counting algorithm as there are no boundary effects at all. This is due to the initial adjustment of the grid over the data.

Due to the dependence of the slope on r, the correct scaling region must be selected manually for every data set. If, for example in the case of a moving analysis in time, D_2 of the different time intervals is to be compared, the possibly different scaling regions must be taken into account – the fluctuations in D_2 obtained from a blind moving analysis might be meaningless. This is also true for an analysis with increasing embedding dimension (see [HS93] for a suggestion to use a Durbin-Watson test to automatically detect "multiscaling").

The computing time of the unmodified Grassberger-Procaccia algorithm increases with the square of the number of data points, a box-assisted modification by Theiler (1987) used in this work reduces the time to the order of $N \log N$, where N is the number of data points. Because this amount of time is frequently still too large, $C(r)$ is usually only determined for a subset of points, the so-called reference points. This leads to further uncertainty in the result: in tests performed during this work, the estimate of D_2 still depended on the particular choice of the randomly selected reference points even when the number of reference points was as high as 80% of the number of data points. Although in the analysed cases these variations were well within the conservative estimate of the error made when using all data points, the dependence of the result on the choice of the reference points (number as well as distribution) should be kept in mind.

The original box-counting method requires too many data points in higher dimensions (roughly 10^d, where d is the expected dimension), too much computer memory and is too slow (obviously, no limitation to a subset of reference points is possible as all points of the set have to be counted).

As the fractal dimension describes how many new pieces of a set are resolved as the resolution scale is increased ([Man83]), the fractal dimension can be estimated by comparing properties between any two scales (as the set is statistically self-similar), i.e.

$$D_0 = \frac{d \log N(r)}{d \log 1/r} = \frac{\log N(r_i)}{\log 1/r_i} - \frac{\log N(r_j)}{\log 1/r_j}, i \neq j.$$

In the practical modified box-counting algorithm, the coordinates of every point x in the d-dimensional set are rescaled to the range $[0, 2^k - 1]$, where k depends on the integer representation of the respective compiler. Then the set is covered by a grid of d-dimensional cubes of edge sizes $2^m, 0 \leq m \leq k$, the "boxes". Next the coordinates are coded binarily according to the box they lie in by a simple binary AND operation. To determine the actual number of points in the boxes of a particular size, the bits of the previously ordered points are scanned. A more detailed description of the modified box-counting algorithm may be found in [LT89].

When r is so small that every point lies in a distinct box, i.e. $N(r) = N$, the function $N(r)$ has saturated and the maximum resolution possible with respect to the data size is reached. These values of $N(r)$, as well as values due to saturation because of the decreasing resolution, should not be used for the slope estimation. To ensure that $N(r)$ is not near saturation, Liebovitch and Toth (1989) exclude values of $N(r)$ for $N(r) \leq N/5$. Likewise the values for $m = k, k - 1$ are discarded. Thus in this case the method of fit is not determined from the shape of the $\log - \log$ plot, but from the properties of the set itself and is always the same. Therefore results are more comparable and the method seems suitable even for an automatic detection of changes in the fractal dimension like in the continuous observation of the fractal properties of seismicity distribution. The method was extended to arbitrary values of q in this work, but, especially for $q < 0$, the selection of the scaling region had to be performed by nonlinear optimisation (cf. Chapter 3).

However, the danger of averaging over real multiscaling behaviour remains. In this work, the algorithm was therefore extended to include statistical output like the standard deviation of the slope, the correlation coefficient of the fitted line with the points in the scaling region and confidence limits obtained from a Students t-test. These values must be monitored carefully if the analysis is carried out using linear regression and performed blindly, i.e. without visual inspection of the log − log plot. More advanced checks are discussed by Gonzato et al. (1998). See also Section 3.4.

Liebovitch and Toth (1989) found the accuracy of their algorithm to be equal or better than the Grassberger-Procaccia algorithm and reached execution times of up to about 36 times faster for 10^4 data points. Similar results were found during numerous tests and analyses throughout this work. At least for dimensions smaller than three, i.e. for example for the analysis of epicentre distributions, the modified box-counting algorithm seems preferably to sphere counting. This is especially true for the automatic detection of changes in the first three generalised dimensions. In this work the algorithm has been extended for the determination of arbitrary D_q, i.e. q may be set freely. For $q < 0$ and $q > 3$, the slopes are determined by nonlinear optimisation as described in Chapter 3 instead of the original linear approach outlined above. For low dimensional data the accuracy of the results frequently surpassed the one of generalised sphere counting (see also [BvBS90] for efficient and accurate determination of generalised dimensions by box-counting).

B. Fractal Cluster Analysis

In a paper by Henderson *et al.* (1994), a fractal analysis of seismicity was carried out after an observation area in north-eastern Brazil had been separated into clusters of distinct seismicity by cluster analysis. The idea is that one should not average the possibly peculiar (fractal) properties of distinct seismogenic zones (cf. also [GC95]). These zones do not have to correlate with e.g. different branches of a fault system or, more generally, any distinct geological features known *a priori*. Although a multifractal analysis disentangles a structure consisting of intertwined fractals of different scaling, it might still make sense to observe the temporal evolution of different zones separately. This is for example because of possible stress transfers between regions, leading to a critical state in one region while another becomes uncritical. Thus a combined cluster and multifractal analysis seems rewarding (the analysis by Henderson *et al.* (1994) was monofractal). No such analysis was performed yet, but a consideration of the problem led to the following idea.

The objective of cluster analysis is to separate data into distinct groups of objects. Basically, objects belonging to the same group should be as similar (close together) as possible, while objects in different groups should be as dissimilar (far apart) as possible. Cluster analysis establishes the groups, whereas discriminant analysis assigns objects to groups previously defined.

In up to three dimensions, subjective classification by visual inspection is possible but not objective as no measure of the quality or ambiguity of the chosen clustering is obtainable. Automatic classification is an independent science discipline and often used in the context of artificial intelligence (AI), particularly in pattern recognition.

A big variety of algorithms for cluster analysis exists (e.g. [KR90]) and many have been applied to seismicity patterns (e.g. [HMPT94]). Although the choice of the actual clustering algorithm depends on the data and purpose, several different algorithms might still be applicable to the same problem—as cluster analysis does not try to prove a preconceived hypothesis but is merely descriptive, the most "pleasing" method may be chosen.

Clustering algorithms may be separated into *partitioning* and *hierarchical* methods. In the case of partitioning, the data (for example a set of n vectors) is separated into k distinctive clusters which do not overlap and whose number is necessarily $\leq n$. k must be estimated from the data (by visual inspection) in advance or it can be determined automatically by trying a range of k and picking the partition which best fits some numerical criterion. This method may be used to find an existing structure or to impose one onto the data.

Hierarchical methods do not partition into k clusters, but their output consists of partitions from $k = 1$ to $k = n$ in one run. During the step from $k = l$ to $k = l + 1$, one of the l clusters splits up into two (divisive method) or two of the $l + 1$ clusters are combined to give l clusters (agglomerative method). Divisive and agglomerative methods yield quite different results. The successively resulting sub-partitions are not necessarily sensible so that the hierarchical method cannot replace partitioning methods. The advantage of hierarchical methods lies in the speed of computation, their disadvantage is that erroneous decisions cannot be corrected. They have been very successfully applied in biology but in the case of earthquake clustering, partitioning methods seem more appropriate.

Partitioning methods are widely applied to find out whether two-dimensional distributions contain groups. The above mentioned applications to earthquakes use a partitioning around medoids: to obtain k clusters, k representative objects or seeds have to be chosen initially. Then all remaining objects are assigned to their nearest representative object. The optimal representative object, where the average distance to all other elements of that cluster is minimised, is called the medoid; if the square of the distances is used, it is called the centroid. This method produces spherical clusters, i.e. might miss elongated groups which occur in the case of earthquakes which frequently happen along elongated structures ("fault segments").

The method chosen here is fuzzy logic cluster analysis (e.g. [KR90, Miy90]). It is very different from other clustering methods and involves no seeds. The computations are comparatively complex and therefore time consuming. Also the amount of output data is bulky. However, the great advantage is that the method does not yield one single hard clustering but so-called membership coefficients for each object and cluster. Thus the output consists of a $K * n$ matrix, where every element denotes the probability of the respective object to belong to the respective cluster.

In this work, the application of fuzzy cluster analysis to small earthquake data sets frequently yielded events with extremely small membership coefficients. The latter indicates that these events cannot be assigned to any well-defined seismogenic structure and should possibly be discarded completely in a fractal analysis. This might not only be physically significant, but also prevents erroneous clusters which would emerge using any of the hard-cluster-methods. A combined fuzzy-cluster multifractal analysis seems rewarding as it might render clearer precursors as were obtained in Section 7.3.1 for example: without separation into clusters, the multifractal spectra possibly only yielded some average information which obscured more localised precursory behaviour.

Acknowledgements

Thanks to: Torao Tanaka for inviting me to Japan and extensive support while there. H.-J. Kümpel for various support, stimulating discussion and friendship. The Japanese Ministry of Education and the German Academic Exchange Service (DAAD) for funding. Kunihiko Watanabe (DPRI) for strain and earthquake data. Yuzo Ishikawa of the Japan Meteorological Agency (JMA) for earthquake data. Masatako Ando for being enthusiastic about fractals and supporting my research. Hiroshi Fukuoka for landslide data and discussions. George Igarashi and Hiroshi Wakita for radon and seismicity data and for discussions at Tokyo University. Keisuke Ito for discussion and support. Tony Roberts for making available source code and for discussions. L. Kaufman for making available source code. Thomas Kruel for making available his Ph.D. thesis along with source codes. Tamas Vicsek and Ingmar Procaccia for discussion and review. Charles Hooge for sending me his M.Sc. thesis and other publications. H. J. Neugebauer for encouraging me to publish this work as a book.

These people took the time for sometimes extensive discussions: Julien C. Sprott, Hideki Takaysu, John Sarraille, John Russ, Gerald A. Edgar, Wolfgang Rabbel. Several people on the Internet, either through NetNews or mailing lists, contributed to this research by discussions or by pointing out resources.

My wife Eden made it all worthwhile and helped me keep my sanity♡

References

[AAD⁺86] N. B. Abraham, A. M. Albano, B. Das, G. de Guzman, S. Yong, R. S. Gioggia, G. P. Puccioni, and J. R. Tredicce. Calculating the dimension of attractors from small data sets. *Phys. Lett. A*, 114:217+, 1986.

[ABST93] H. D. I. Abarbanel, R. Brown, J. L. Sidorowich, and L. Sh. Tsimring. The analysis of observed chaotic data in physical systems. *Rev. Mod. Phys.*, 65:1331–1392, 1993.

[AFH94] J. Argyris, G. Faust, and M. Haase. *An exploration of chaos*. North-Holland, Amsterdam, 1994.

[Aki81] K. Aki. A probabilistic synthesis of precursory phenomena. In D. W. Simpson and P. G. Richards, editors, *Earthquake Prediction: An International Review*, number 4 in Maurice Ewing Ser., pages 566+. AGU, Washington, DC, 1981.

[AL76] A. A. Anis and E. H. Lloyd. The expected value of the adjusted rescaled hurst range of independant normal summands. *Biometrika*, 1(63):111+, 1976.

[ASB87] C. A. Aviles, C. H. Scholz, and J. Boatwright. Fractal analysis applied to characteristic segments of the San Andreas Fault. *J. Geophys. Res.*, 92:331+, 1987.

[Bak86] P. Bak. The devil's staircase. *Physics Today*, pages 38+, 1986.

[BAL⁺94] A. Beghdadi, C. Andraud, J. Lafait, J. Peiro, and M. Perreau. Entropic and multifractal analysis of disordered morphologies. In T. Vicsek, M. Shlesinger, and M. Matsushita, editors, *Fractals in Natural Science*, pages 360+, Singapore, 1994. World Scientific. This is a full INPROCEDINGS entry.

[Bar88] M. F. Barnsley. *Fractals everywhere*. Academic Press, San Diego, 1988.

[BGT87] J. Bebién, C. Gagny, and S. S. Tanani. Les associations de magmas acides et basiques: des objects fractals? *C. R. Acad. Sci. Paris*, 305:277+, 1987.

[Ble91] T. G. Blenkinsop. Cataclasis and processes of particle-size reduction. *PAGEOPH*, 136:59+, 1991.

[BMPV93] S. Borgani, G. Murante, A. Provenzale, and R. Valdarnini. Multifractal analysis of the galaxy distribution: Reliability of results from finite data sets. *Phys. Rev. E*, 47(6):3879+, 1993.

[Bod93] B. Bodri. A fractal model for regional seismicity at Izu Peninsula, Japan. In *Fractals in Natural Sciences: Int. Conference on the Complex Geometry in Nature*, page E2, Budapest, 1993. Book of Abstracts, personal communication.

[Boe88] D. C. Boes. Schemes exhibiting hurst behaviour. In J. N. Srivastava, editor, *Probability and Statistics. essays in Honour of F. A. Graybill*, pages 21+. Elsevier Science Publishers, Amsterdam, 1988.

[BPPV84] R. Benzi, G. Paladin, G. Parisi, and A. Vulpiani. On the multifractal nature of fully developed turbulence and chaotic system. *J. Phys.*, 18:3521, 1984.

[BS85]   S. R. Brown and C. H. Scholz. Broad bandwith study of the topography of natural rock surfaces. *J. Geophys. Res.*, 90:12575+, 1985.

[BS93]   C. Beck and F. Schlögel. *Thermodynamics of Chaotic Systems.* Cambridge Nonlinear Science Series. Cambridge University Press, Cambridge, New York, Melbourne, 1993.

[BvBS90] A. Block, W. von Bloh, and H. J. Schellnhuber. Efficient box-counting determination of generalised fractal dimensions. *Phys. Rev. A*, 42(4):1869+, 1990.

[CE91]   M. Casdagli and S. Eubank. Nonlinear modeling and forecasting. In *Proceedings of the NATO/ Santa Fe Institute conference on nonlinear forecasting and modeling, September 1990*, volume XI, Reading Mass., 1991. Addison-Wesley.

[CJ89a] A. Chhabra and R. V. Jensen. Direct determination of the $f(\alpha)$ singularity spectrum. *Phys.Rev.Lett.*, 69:1327+, 1989.

[CJ89b] D. J. Crossley and O. G. Jensen. Fractal velocity models in refraction seismology. *PAGEOPH*, 131(1-2):61+, 1989.

[CJV92] A. Crisanti, M. H. Jensen, and A. Vulpiani. Strongly intermittent chaos and scaling in an earthquake model. *Phys. Rev. E*, 46(12):7363+, 1992.

[CLS94] J. M. Carlson, J. S. Langer, and B. E. Shaw. Dynamics of earthquake faults. *Reviews of Modern Physics*, 66(2):657+, 1994.

[Cro86] M. J. Crozier. *Landslides: Causes, consequences and environment.* Croom Helm, 1986.

[Cut93] C. Cutler. A review of the theory and estimation of fractal dimensions. In H. Tong, editor, *Nonlinear Time series and Chaos*, pages 566+. World Scientific, Singapore, 1993.

[EER95] EERI. The Hyogo-Ken Nanbu Earthquake. Technical report, Earthquake Engineering Research Institute, Oakland, 1995.

[EKRC86] J.-P. Eckmann, S. O. Kamphorst, D. Ruelle, and S. Ciliberto. Liapunov exponents from time series. *Phys. Rev. A*, 34:4971–4979, 1986.

[EM92]   C. J. G. Evertsz and B. B. Mandelbrot. Multifractal measures. In H. O. Peitgen, H. Jürgens, and D. Saupe, editors, *Chaos and Fractals*, pages 921+. Springer, New York, 1992.

[EQE95] EQE. The January 17, 1995 Kobe Earthquake. Technical report, EQE, San Francisco, 1995. info@eqe.com.

[ER94]   L. M. Emmerson and A. J. Roberts. Fractal and multi-fractal patterns of seaweed settlement. made available through public ftp server ftp.usq.edu.au, April 1994.

[Eve70] J. F. Evernden. Study of regional seismicity and associated problems. *Seis. Soc. Am. Bull.*, 60:393+, 1970.

[Fed88] J. Feder. *Fractals*. Physics of Solids and Liquids. Plenum Press, New York, London, 1988.

[FHG94] H. Fukuoka, H. Hiura, and C. Goltz. Fractal aspects of the landslide distribution and size-frequency relation of landslides in hokkaido. In *Proc. Annual. Conf. of the Japanese Landslide Society*, pages 23+, 1994. in Japanese.

[FPH90] M. E. Farrell, A. Passamante, and T. Hediger. Comparing a nearest-neighbor estimator of local attractor dimensions for noisy data to the correlation dimension. *Phys. Rev. A*, 41(12):6591+, 1990.

[FS89]   R. H. Fluegeman and R. S. Snow. Fractal analysis of long-range paleocli-
          matic data: Oxygen isotope record of pacific core v28-239. *PAGEOPH*,
          131(1-2):307+, 1989.

[FSN78]   U. Frisch, P. Sulem, and M. Nelkin. A simple dynamical model of inter-
          mittent fully developed turbulence. *J. Fluid. Mech.*, 87:719+, 1978.

[GC95]   C. Godano and V. Caruso. Multifractal analysis of earthquake cata-
          logues. *Geophys. J. Int.*, 121:385+, 1995.

[Gel97]   R. J. Geller. Earthquake prediction: a critical review. *Geophys. J. Int.*,
          131:425+, 1997.

[GGP90]   M. B. Geilikman, T. V. Golubeva, and V. F. Pisarenko. Multifractal
          patterns of seismicity. *Earth and Planetary Science Letters*, 99:127+,
          1990.

[GJKM97] R. J. Geller, David D. Jackson, Yan Y. Kagan, and Francesco Mulargia.
          Earthquakes cannot be predicted. *Science*, 275:1616+, 1997.

[GMM98]   G. Gonzato, F. Mulargia, and W. Marzocchi. Practical application of
          fractal analysis: problems and solutions. *Geophys. J. Int.*, 132:275+,
          1998.

[GMP98]   J. B. Gómez, Y. Moreno, and A. F. Pacheco. Probabilistic approach
          to time-dependent load-transfer models of fracture. *To appear in Phys.
          Rev. E*, 1998.

[Gol90]   C. Goltz. Realisierung einer mittelfristigen Echtzeit-Wasserstandsvor-
          hersage für die Deutsche Bucht am Beispiel des Pegels Büsum. Diplo-
          marbeit, Christian-Albrechts-Universität zu Kiel, 1990.

[Gol96]   C. Goltz. Multifractal and entropic properties of landslides in Japan.
          Geolog. Rundsch., 85:71+, 1996.

[Gol97]   C. Goltz. Using determinism in earthquake inter-arrival times to look for
          possible precursory behaviour. In *AGU 1997 Fall Meeting*, volume 46,
          page F478, Washington, November 1997. Supplement to Eos, Transac-
          tions, AGU.

[GP83]   P. Grassberger and I. Procaccia. Characterization of strange attractors.
          Phys. Rev. Lett., 20:346+, 1983.

[GP84]   P. Grassberger and I. Procaccia. Dimensions and entropies of strange
          attractors from a fluctuating dynamics approach. *Physica D*, 13:34+,
          1984.

[Gra88]   P. Grassberger. Finite sample corrections to entropy and dimension
          estimates. *Phys. Letts. A*, 128:369+, 1988.

[Gra94]   P. Grassberger. Efficient large-scale simulations of a uniformly driven
          system. *Phy. Rev. E*, 49(4):2436+, 1994.

[GW88]   C. Goltz and W. Welle. Iterative Funktionensysteme: Eine neue Meth-
          ode in der Computergraphik. *Journal der Deutschen Geophysikalischen
          Gesellschaft*, 4:24+, 1988. in German.

[GWSP82] H. S. Greenside, A. Wolf, J. Swift, and T. Pignataro. Impracticality of
          a box-counting algorithm for calculating the dimensionality of strange
          attrators. *Phys. Rev. A*, 25:3453+, 1982.

[Hai93]   J. Haikun. The multifractal local scaling feature of spatial 'energy gener-
          ating' and its seismic precursor information. In *Int. Symp. on Fractals
          and Dyn. Sys. in Geoscience*, volume 1, pages 12+, Frankfurt, April 1993.
          Book of Abstracts.

[Hau19]   F. Hausdorff. Dimension und äusseres Mass. *Mathematische Annalen*,
          79:157+, 1919.

[HC91]   P. Hubert and J. P. Carbonel. Fractal characterization of intertropical precipitations variability and anisotropy. In D. Schertzer and S. Lovejoy, editors, *Non-Linear Variability in Geophysics*, pages 209+. Kluwer Academic Publishers, Dordrecht, 1991.

[HE89]   J. W. Havstad and C. L. Ehlers. Attractor dimension of non-stationary dynamical sytems from small data sets. *Phys. Rev. A*, 39:845+, 1989.

[Hew86]   T. A. Hewett. Fractal distributions of reservoir heterogeneity and their influence on fluid transport. In *61st Annu. SPE Tech. Conf. Pap.* SPE 15386, New Orleans, 1986.

[HF94]   H. Hiura and H. Fukuoka. Fractal distribution characteristics of landslides in hokkaido isl., sikoku isl. and tohoku district. In *East Asia Symposium and Field Workshop on Landslides and Debris Flow*, pages 35+, 1994.

[HH94]   S.-Z. Hong and S.-M. Hong. An amendment to the fundamental limits on dimension calculations. *Fractals*, 2(1):123+, 1994.

[HI91]   T. Hirata and M. Imoto. Multifractal analysis of spatial distribution of microeathquakes in the Kanto region. *Geophys. J. Int.*, 107:155+, 1991.

[Hir87a] T. Hirata. Omori's power law aftershock sequences of microfracturing in rock fracture experiment. *J. Geophys. Res.*, 92:6215+, 1987.

[Hir87b] T. Hirata. A timeseries of AE events of Andesite under the triaxial compression. In *Proc. 7th Japan Symp. on Rock Mech.*, pages 301+, 1987.

[Hir89a] T. Hirata. A correlation between the *b* value and the fractal dimension of earthquakes. *J. Geophys. Res.*, 94:7507+, 1989.

[Hir89b] T. Hirata. Fractal dimension of fault systems in Japan: Fractal structure in rock fracture geometry at various scales. *PAGEOPH*, 131:157+, 1989.

[HIY92]   T. Hirabayashi, K. Ito, and T. Yoshii. Multifractal analysis of earthquakes. *PAGEOPH*, 138(4):591+, 1992.

[HJK⁺86] T. C. Halsey, M. H. Jensen, L. P. Kadanoff, I. Procaccia, and B. Shraiman. Fractal measures and their singularities: The characterization of strange sets. *Phys. Rev. A*, 33:1141+, 1986.

[HK79]   T. C. Hanks and H. Kanamori. A moment-magnitude scale. *J. Geophys. Res.*, 84:2348+, 1979.

[HLS⁺94] C. Hooge, S. Lovejoy, D. Schertzer, S. Pecknold, J.-F. Malouin, and F. Schmitt. Multifractal phase transitions: The origin of self-organized criticality in earthquakes. *Nonlinear Processes in Geophysics*, 1:191+, 1994.

[HMPT94] J. Henderson, I. G. Main, R. G. Pearce, and M. Takeya. Seismicity in north-eastern brazil: fractal clustering and the evolution of the *b* value. *Geophys. J. Int.*, 116:217+, 1994.

[Hoo93]   C. Hooge. Earthquakes as a space-time multifractal process. MSc thesis, McGill University, 1993.

[HP83]   H. G. E. Hentschel and I. Procaccia. The infite number of generalized dimensions of fractals and strange attractors. *Physica D*, 8:435+, 1983.

[HS93]   H. M. Hastings and G. Sugihara. *Fractals-A Users's Guide for the Natural Sciences*. Oxford University Press, Oxford, 1993.

[HSI87]   T. Hirata, T. Satoh, and K. Ito. Fractal structure of spatial distribution of microfracturing in rock. *Geophys. J. Roy. Astr. Soc.*, 90:369+, 1987.

[Hsü92]   K. J. Hsü. Fractal geometry of global change in earth history. In *29th International Geological Congress*, volume 1, page 10, Tsukuba, August 1992. Book of Abstracts.

[HT90]   J. Huang and D. L. Turcotte. Are earthquakes an example of deterministic chaos? *Geophys. Res. Lett.*, 17:223+, 1990.

[HT92]   J. Huang and D. L. Turcotte. Chaotic seismic faulting with a mass-spring model and velocity-weakening friction. *Pure Appl. Geophys.*, 138:569+, 1992.

[Hur51]   H. E. Hurst. Long-term storage capacity of reservoirs. *Trans. Am. Soc. Civil Eng.*, 116:770+, 1951.

[Hur56]   H. E. Hurst. Methods of using long-term storage in reservoirs. *Proc. Inst. Civil Eng.*, 5(Part 1):519+, 1956.

[Ito92]   K. Ito. Towards a new view of earthquake phenomena. *PAGEOPH*, 138:531+, 1992.

[IW90]   G. Igarashi and H. Wakita. Groundwater radon anomalies associated with earthquakes. *Tectonophysics*, 180:237+, 1990.

[KA75]   H. Kanamori and D. L. Anderson. Theoretical basis of some empirical relations in seismology. *Seis. Soc. Am. Bull*, 65:1073+, 1975.

[Kag92]   Y. Y. Kagan. Seismicity: Turbulence of solids. *Non-Linear Science Today*, 2:8+, 1992.

[Kag93]   Y. Y. Kagan. Statistics of characteristic earthquakes. *Bull. Seis. Soc. Am.*, 83:22+, 1993.

[Kan78]   H. Kanamori. Quantification of earthquakes. *Nature*, 271:411+, 1978.

[Kat95]   H. Katao. personal communication, October 1995. Dr. Katao's e-mail address is katao@epdpri1.dpri.kyoto-u.ac.jp.

[KG95]   D. Kaplan and L. Glass. *Understanding nonlinear dynamics.* Springer Verlag, New York, 1995.

[KI92]   M. B. Kennel and S. Isabelle. Method to distinguish chaos from colored noise and to determine embedding parameters. *Phys. Rev. A*, 46(6):3111+, 1992.

[Kin83]   G. King. The accomodation of large strains in the upper lithosphere of the earth and other solids by self-similar fault systems: The geometrical origin of the *b*-value. *PAGEOPH*, 121:761+, 1983.

[KK78]   Y. Y. Kagan and L. Knopoff. Statistical study of the occurence of shallow earthquakes. *Geophys. J. Roy. Astr. Soc.*, 55:55+, 1978.

[KK80]   Y. Y. Kagan and L. Knopoff. Spatial distribution of earthquakes: the two-point correlation function. *Geophys. J. Roy. Astr. Soc.*, 62:303+, 1980.

[Kle74]   V. Klemens. The hurst phenomenon: A puzzle? *Water Resour. Res.*, 10(4):675+, 1974.

[Kno64]   L. Knopoff. Earth tides as a triggering mechnism for earthquakes. *Seism. Soc. Am. Bull.*, 54:1865+, 1964.

[Kol59]   A. N. Kolmogorov. Entropy per unit time as a metric invariant of automorphisms. *Dokl. Akad. Nauk. SSSR*, 124:754+, 1959. (English translation in Math. Review., **21**, 2035+).

[Kom95]   S. Komatsu. *Japan Sinks.* Kodansha, Tokyo, 1995.

[Kor92]   G. Korvin. *Fractal Models in the Earth Sciences.* Elsevier, Amsterdam, 1992.

[KR90]   L. Kaufman and P. J. Rousseeuw. *Finding groups in data: an introduction to cluster analysis.* John Wiley & Sons, New York, 1990.

[Kru91]   Th.-M. Kruel. Scount: A program to calculate the correlation and information dimension from attractors by the method of "sphere-counting". public ftp-server ftp.phys-chemie.uni-wuerzburg.de, October 1991. Obtained with other programs and documentation for nonlinear analysis.

[Kru92]   T.-M. Kruel. *Zeitreihenanalyse nichtlinearer Systeme: Chaos und Rauschen.* PhD dissertation, Bayerische Julius-Maximilian-Universität Würzburg, Würzburg, 1992. (in German).

[Küm91]   H.-J. Kümpel. Hydrologic and geochemical precursors: Implications for crustal models. In *Proceed. Int. Conf. on Earthquake Prediction—State of the Art*, pages 249+, Strasbourg, October 1991.

[LE98]   K. Lehnertz and C. E. Elger. Can epileptic seizures be predicted? Evidence from nonlinear time series analysis of brain electrical activity. *Phys. Rev. Lett.*, 80(22):5019+, 1998.

[Lom94]   C. Lomnitz. *Fundamentals of Earthquake Prediction*. John Wiley & Sons, New York, 1994.

[Lor63]   E. N. Lorenz. Deterministic nonperiodic flow. *J. Atmos. Sci.*, 20:130+, 1963.

[LT89]   S. Liebovitch and T. Toth. A fast algorithm to determine fractal dimensions by box counting. *Phys. Lett. A*, 141:386+, 1989.

[Mai96]   Ian Main. Statistical physics, seismogenesis and seismic hazard. *Reviews of Geophysics*, 4(34):433+, 1996.

[Man65]   B. B. Mandelbrot. Une classe de processes stochastique homothetique a soi; application a la loi climatologique de h. e. hurst. *Comptes Rendus Acad. Sci. Paris*, 260:3274+, 1965.

[Man77]   B. B. Mandelbrot. *Fractals: Form, Chance and Dimension*. W. H. Freeman and Company, New York, 1977.

[Man83]   B. B. Mandelbrot. *The Fractal Geometry of Nature*. W. H. Freeman and Company, New York, 1983. Updated and Augmented Edition.

[McC94]   J. L. McCauley. *Chaos, Dynamics and Fractals – an algorithmic approach to deterministic chaos*. Cambridge Nonlinear Science Series. Cambridge University Press, Cambridge, New York, Melbourne, 1994. First paperback edition with corrections.

[Mei94]   R. Meissner. Non-linear processes in earthquake prediction research, a review. In J. H. Kruhl, editor, *Fractals and Dynamic Systems in Geoscience*, pages 159+. Springer Verlag, Heidelberg, 1994.

[Miy90]   S. Miyamoto. *Fuzzy sets in information retrieval and cluster analysis*. Kluwer Academic Publishers, Dordrecht, 1990.

[MM70]   B. B. Mandelbrot and K. McCamy. On the secular pole motion and the chandler wobble. *Geophys. J. R. Astr. Soc.*, 42(21):217+, 1970.

[MN68]   B. B. Mandelbrot and J. W. Van Ness. Fractional brownian motions, fractional noises and applications. *SIAM Rev.*, 10(4):422+, 1968.

[Moo92]   F. C. Moon. *Chaotic and Fractal Dynamics*. Plenum Press, New York, London, 1992.

[MS87]   C. Meneveau and K. R. Sreenivasan. The multifractal spectrum of the dissipation field in turbulent flow. In M. D. Van and B. Nicolis, editors, *Physics of Chaos and Far From Equilibrium*. North-Holland, Amsterdam, 1987.

[MT91]   M. Matsuzaki and H. Takayasu. Fractal features of the earthquake phenomenon and a simple mechanical model. *J. Geophys. Res.*, 96(B12):19925+, 1991.

[MW68]   B. B. Mandelbrot and J. R. Wallis. Noah, Joseph, and the operational hydrology. *Water Resour. Res.*, 4(5):909+, 1968.

[MW69]   B. B. Mandelbrot and J. R. Wallis. Some long-run properties of geophysical records. *Water Resour. Res.*, 5(2):321+, 1969.

[Nak90]   H. Nakanishi. Cellular-automaton model of earthquakes with deterministic dynamics. *Phys. Rev. A*, 41(12):7086+, 1990.

[NW77]   M. Noguchi and H. Wakita. A method for continuous measurement of radon in groundwater for earthquake prediction. *J. Geophys. Res.*, 82:1353+, 1977.

[OA87]   P. G. Okubo and K. Aki. Fractal geometry in the San Andreas Fault System. *J. Geophys. Res.*, 92:345+, 1987.

[OA91]   Y. Ogata and K. Abe. Some statistical features of the long-term variation of the global and regional seismicity. *Int. Stat. Review*, 59:139+, 1991.

[OFC92]   Z. Olami, H. J. S. Feder, and K. Christensen. Self-organized criticality in a continuous, nonconservative cellular automaton modeling earthquakes. *Phys. Rev. Lett.*, 68(8):1244+, 1992.

[Oga88]   Y. Ogata. Statistical models for earthquake occurrences and residual analysis for point processes. *J. Am. Stat. Assoc.*, 83:401+, 1988.

[OGY89]   E. Ott, C. Grebogi, and J. A. Yorke. Theory of first order phase transitions for chaotic attractors of nonlinear dynamical systems. *Phys. Lett. A*, 135:334+, 1989.

[OM92]   S. Ouchi and M. Matsushita. Measurement of self-affinity on surfaces as a trial application of fractal geometry to landform analysis. *Geomorphology*, 5:115+, 1992.

[Ott93]   E. Ott. *Chaos in Dynamical Systems.* Cambridge University Press, New York, 1993.

[OU86]   T. Ouchi and T. Uekawa. Statistical analysis of the spatial distribution of earthquakes - variation of the spatial distribution of earthquakes before and after large earthquakes. *Phys. Earth Planet Int.*, 44:211+, 1986.

[P+92]   W. H. Press et al. *Numerical Recipes in C.* Cambridge University Press, Cambridge, 1992. Second Edition.

[PBJD79] A. W. Pettis, T. A. Bailey, A. K. Jain, and R. C. Dubes. An intrinsic dimensionality estimator from near-neighbor information. *IEEE Transactions on Pattern Analysis and Machine Intelligence*, PAM-1(1):25+, 1979.

[PJ94]   P. Puster and T. H. Jordan. Stochastic analysis of mantle convection experiments using two-point correlation functions. *Geophys. Res. Lett.*, 21(4):305+, 1994.

[PS87]   K. Pawelzik and H. G. Schuster. Generalized dimensions and entropies from a measured time series. *Phys. Rev. A*, 35:481, 1987.

[PS88]   H.-O. Peitgen and D. Saupe, editors. *The Science of Fractal Images.* Springer, Heidelberg, 1988.

[PVBV93] A. Provenzale, B. Villone, A. Babiano, and R. Vio. Intermittency, phase randomization and generalized fractal dimensions. *Journal of Bifurcation and Chaos*, 3:729+, 1993.

[Ric58]   C. F. Richter. *Elementary Seismology.* Freeman and Co., San Francisco, 1958.

[RKG96]   J. B. Rundle, W. Klein, and S. Gross. Dynamics of a traveling density wave model for earthquakes. *Phys. Rev. Lett.*, 76:4285+, 1996.

[Rob91]   D. A. Roberts. Is there a strange attractor in the magnetosphere. *J. Geophys. Res. A*, 96(9):16031+, 1991.

[RSSS93] P. A. Rydelek, I. Selwyn-Sacks, and R. Scarpa. On tidal triggering of earthquakes at Campi Flegrei, Italy. *Geophys. J. Int.*, 109:125+, 1993.

[Rus90]   J. C. Russ. Surface characterization: Fractal dimensions, hurst coefficients and frequency transforms. *Journal of Computer Assisted Microscopy*, 2(3):161+, 1990.

[Rus94]   J. C. Russ. *Fractal Surfaces.* Plenum Press, New York, 1994.

[RY90]   J. B. Ramsey and H. J. Yuan. The statistical properties of dimension calculations using small data sets. *Nonlinearity*, 3:155+, 1990.

[S+84]   M. A. Sadovskiy et al. Characteristic dimensions of rock and hierarchical properties of seismicity. *Izvestiya Acad. Sci. USSR, Phys. Solid Earth*, 20:87+, 1984.

[SA86]   C. H. Scholz and C. A. Aviles. The fractal geometry of faults and faulting. In S. Das, J. Boatwright, and C. H. Scholz, editors, *Earthquake Source Mechanics*, number 6 in Maurice Ewing Ser., pages 147+. AGU, 1986.

[SAK88]   M. Suzuki, T. Asakawa, and S. Kobayashi. Examination of critical rainfall for landslides with rain fall radar information – the case of typhoon no. 10, 1983 attack on hyogo and kyoto. In *Proceed. Ann. Conf. of Erosion Control Society*, pages 81+, 1988. in Japanese.

[Sar92]   J. J. Sarraille. Developing algorithms for measuring fractal dimensions. public ftp-server csustan.csustan.edu, June 1992. Obtained with other documentation of program FD3.

[Sar93]   J. J. Sarraille. personal communication, October 1993. Prof. Sarraille's e-mail address is john-s@u-aizu.ac.jp.

[Sat88]   H. Sato. Fractal interpretation of the linear relation between logarithms of maximum amplitude and hypocentral distance. *Geophys. Res. Let.*, 15:373+, 1988.

[SB89]   C. G. Sammis and R. L. Biegel. Fractals, fault-gouge and friction. *PAGEOPH*, 131:255+, 1989.

[Sch68]   C. H. Scholz. Microfractures, aftershocks and seismicity. *Bull. seism. Soc. Am.*, 58:1117+, 1968.

[Sch88]   H. G. Schuster. *Deterministic chaos: An introduction*. VCH, Weinheim, 1988.

[Sch89]   C. H. Scholz. Global perspectives of chaos. *Nature*, 338:459+, 1989.

[SCL92]   B. E. Shaw, J. M. Carlson, and J. S. Langer. Patterns of activity preceeding large earthquakes. *J. Geophys. Research.*, 97:479+, 1992.

[SD95]   E. Segre and C. Deangeli. Cellular automaton for realistic modelling of landslides. *Nonlinear Proc. Geophys.*, 2:1+, 1995.

[Sha81]   R. Shaw. Strange attractors, chaotic behaviour and information flow. *Z. Naturforsch. A*, 36:80+, 1981.

[SL93]   D. Schertzer and S. Lovejoy. Nonlinear variability in geophysics: Scaling and multifractal processes. Lecture Notes, 1993. AGU Chapman/EGS Richardson Memorial Conference.

[SM88]   R. Stoop and P. F. Meier. Evaluation of lyapunov exponents and scaling functions from time series. *J. Opt. Soc. Am. B*, 5:1037–1045, 1988.

[Smi88]   L. A. Smith. Intrinsic limits on dimension calculations. *Phys. Lett. A*, 133:283+, 1988.

[SP93]   S. Sadovskii and V. F. Pisarenko. Prediction of time series. In Y. A. Kravtsov, editor, *Limits of predictability*, pages 161+. Springer Verlag, Berlin, 1993. Springer series in synergetics.

[Spa82]   C. Sparrow. *The Lorenz equations : Bifurcations, chaos, and strange attractors*. Springer, 1982.

[Spr93]   J. C. Sprott. personal communication, October 1993. Prof. Sprott's e-mail address is sprott@juno.physics.wisc.edu.

[SR95]   J. C. Sprott and G. Rowlands. *Chaos Data Analyzer: The professional version*. American Institute of Physics, New York, 1995. Part of the series *Physics Academic Software*, Editor J. S. Risley, Prerelease Version.

[SVP93]   A. S. Sharma, D. Vassiliadis, and K. Papadopoulos. Reconstruction of low-dimensional magnetospheric dynamics by singular spectrum analysis. *Geophys. Res. Let.*, 20(5):335+, 1993.

[Tak80]   F. Takens. Detecting strange attractors in turbulence. In Rand D. A. and Young L.-S., editors, *Dynamical Systems and Turbulence (Warwick 1980) (Lecture Notes in Mathematics)*, volume 898, pages 366+. Springer, Berlin, 1980.

[Tak90]   H. Takayasu. *Fractals in the Physical Sciences.* Nonlinear Science: Theory and Applications. Manchester University Press, Manchester, New York, 1990.

[Tak91]   T. Takahashi. *Debris Flow.* Belkema, 1991.

[TE93]   J. Theiler and S. Eubank. Don't bleach chaotic data. *Chaos*, 3:771+, 1993.

[TEL+92]   J. Theiler, S. Eubank, A. Longtin, B. Galdrikian, and J. D. Farmer. Testing for nonlinearity in time series: the method of surrogate data. *Physica D*, 58:77+, 1992.

[The87]   J. Theiler. Efficient algorithm for estimating the correlation dimension from a set of discrete points. *Phys. Rev. A*, 36:4456+, 1987.

[The91]   J. Theiler. Some comments on the correlation dimension of 1/f-alpha noise. *Phys. Lett. A*, 155:480+, 1991.

[Tso92]   A. A. Tsonis. *Chaos: From theory to applications.* Plenum, New York, 1992.

[Tur92]   D. L. Turcotte. *Fractals and Chaos in Geology and Geophysics.* Cambridge University Press, Cambridge, 1992.

[Tur97]   D. L. Turcotte. *Fractals and Chaos in Geology and Geophysics.* Cambridge University Press, Cambridge, 1997. Second Edition.

[vdWS88]   W. van de Water and P. Schram. Generalized dimensions from near-neighbor information. *Phys. Rev. A*, 37(8):3118+, 1988.

[Vic92]   T. Vicsek. *Fractal Growth Phenomena, 2nd ed.* World Scientific Publishing Co., Singapore, 1992.

[Vos88]   R. F. Voss. Fractals in nature: From characterization to simulation. In H.-O. Peitgen and O. Saupe, editors, *The Science of Fractal Images*, pages 21+. Springer Verlag, Heidelberg, 1988.

[Wak82]   H. Wakita. Changes in groundwater level and chemical composition. In T. Asada, editor, *Earthquake Prediction Techniques*, pages 175+. University of Tokyo Press, Tokyo, 1982.

[Wat91a]   K. Watanabe. Strain variations of the Yamasaki fault zone, Southwest Japan, derived from extensometer observations. *Bull. of the Dis. Prev. Res. Inst. Kyoto Univ.*, 41(355):53+, 1991. Part 1: On the long-term strain variations....

[Wat91b]   K. Watanabe. Strain variations of the Yamasaki fault zone, Southwest Japan, derived from extensometer observations. *Bull. of the Dis. Prev. Res. Inst. Kyoto Univ.*, 41(354):29+, 1991. Part 2: On the short-term strain variations derived from strain steps....

[WBPP93]   R. Wayland, D. Bromley, D. Pickett, and A. Passamante. Recognizing determinism in a time series. *Phys. Rev. Lett.*, 70:500+, 1993.

[WIN91]   H. Wakita, G. Igarashi, and K. Notsu. An anomalous radon decrease in groundwater prior to an m6.0 earthquake: a possible precursor? *Geophys. Res. Lett.*, 18(4):629+, 1991.

[WNS88]   H. Wakita, Y. Nakamura, and Y. Sano. Short-term and intermediate-term geochemical precursors. *PAGEOPH*, 126:267+, 1988.

[WNS94]   C. Watts, D. E. Newman, and J. C. Sprott. Chaos in reversed-field-pinch plasma simulation and experiment. *Physical Review E*, 49(3):2291+, 1994.

[WSSV85]   A. Wolf, J. B. Swift, H. L. Swinney, and J. A. Vastano. Determining lyapunov exponents from a time series. *Physica D*, 16:285–317, 1985.

[Wys97]   M. Wyss. Cannot earthquakes be predicted? *Science*, 278:487+, 1997.

Index

Springer
and the
environment

At Springer we firmly believe that an
international science publisher has a
special obligation to the environment,
and our corporate policies consistently
reflect this conviction.
We also expect our business partners –
paper mills, printers, packaging
manufacturers, etc. – to commit
themselves to using materials and
production processes that do not harm
the environment. The paper in this
book is made from low- or no-chlorine
pulp and is acid free, in conformance
with international standards for paper
permanency.

Lecture Notes in Earth Sciences